238
Topics in Current Chemistry

Topics in Current Chemistry

Recently Published and Forthcoming Volumes

Natural Product Synthesis II
Volume Editor: Mulzer, J.H.
Vol. 244, 2004

Natural Product Synthesis I
Volume Editor: Mulzer, J.H.
Vol. 243, 2004

Immobilized Catalysts
Volume Editor: Kirschning, A.
Vol. 242, 2004

Transition Metal and Rare Earth Compounds III
Volume Editor: Yersin, H.
Vol. 241, 2004

The Chemistry of Pheromones and Other Semiochemicals II
Volume Editor: Schulz, S.
Vol. 240, 2004

The Chemistry of Pheromones and Other Semiochemicals I
Volume Editor: Schulz, S.
Vol. 239, 2004

Orotidine Monophosphate Decarboxylase
Volume Editors: Lee, J.K., Tantillo, D.J.
Vol. 238, 2004

Long-Range Charge Transfer in DNA II
Volume Editor: Schuster, G.B.
Vol. 237, 2004

Long-Range Charge Transfer in DNA I
Volume Editor: Schuster, G.B.
Vol. 236, 2004

Spin Crossover in Transition Metal Compounds III
Volume Editors: Gütlich, P., Goodwin, H.A.
Vol. 235, 2004

Spin Crossover in Transition Metal Compounds II
Volume Editors: Gütlich, P., Goodwin, H.A.
Vol. 234, 2004

Spin Crossover in Transition Metal Compounds I
Volume Editors: Gütlich, P., Goodwin, H.A.
Vol. 233, 2004

New Aspects in Phosphorus Chemistry IV
Volume Editor: Majoral, J.-P.
Vol. 232, 2004

Elemental Sulfur and Sulfur-Rich Compounds II
Volume Editor: Steudel, R.
Vol. 231, 2003

Elemental Sulfur and Sulfur-Rich Compounds I
Volume Editor: Steudel, R.
Vol. 230, 2003

New Aspects in Phosphorus Chemistry III
Volume Editor: Majoral, J.-P.
Vol. 229, 2003

Dendrimers V
Volume Editors: Schalley, C.A., Vögtle, F.
Vol. 228, 2003

Colloid Chemistry II
Volume Editor: Antonietti, M.
Vol. 227, 2003

Colloid Chemistry I
Volume Editor: Antonietti, M.
Vol. 226, 2003

Modern Mass Spectrometry
Volume Editor: Schalley, C.A.
Vol. 225, 2003

Hypervalent Iodine Chemistry
Volume Editor: Wirth, T.
Vol. 224, 2003

Orotidine Monophosphate Decarboxylase

A Mechanistic Dialogue

Volume Editors: J.K. Lee, D.J. Tantillo

With contributions by
M.R.A. Blomberg · K.L. Byun · J. Gao · K.N. Houk · Y. Hu
R. Kluger · M. Lundberg · B.G. Miller · E.F. Pai · P.E.M. Siegbahn
J.A. Smiley · C. Stanton · D.J. Tantillo · N. Wu

Springer

The series *Topics in Current Chemistry* presents critical reviews of the present and future trends in modern chemical research. The scope of coverage includes all areas of chemical science including the interfaces with related disciplines such as biology, medicine and materials science. The goal of each thematic volume is to give the non specialist reader, whether at the university or in industry, a comprehensive overview of an area where new insights are emerging that are of interest to a larger scientific audience.

As a rule, contributions are specially commissioned. The editors and publishers will, however, always be pleased to receive suggestions and supplementary information. Papers are accepted for *Topics in Current Chemistry* in English.

In references *Topics in Current Chemistry* is abbreviated Top Curr Chem and is cited as a journal.

Visit the TCC content at http://www.springerlink.com/

ISSN 0340-1022
ISBN 3-540-20566-7
DOI 10.1007/b84246
Springer-Verlag Berlin Heidelberg New York

Library of Congress Control Number: 2004106904

Springer-Verlag is a part of Springer Science+Business Media

springeronline.com

Cover design: KünkelLopka, Heidelberg/design & production GmbH, Heidelberg
Typesetting: Stürtz AG, Würzburg

02/3020 ra – 5 4 3 2 1 0 – Printed on acid-free paper

Volume Editor

Prof. Jeehiun K. Lee
Department of Chemistry
and Chemical Biology
Rutgers University
Piscataway, NJ 08854, USA
jklee@rutchem.rutgers.edu

Prof. Dean J. Tantillo
Department of Chemistry
University of California
Davis, CA 95616, USA
tantillo@chem.ucdavis.edu

Editorial Board

Topics in Current Chemistry Also Available Electronically

For all customers who have a standing order to Topics in Current Chemistry, we offer the electronic version via SpringerLink free of charge. Please contact your librarian who can receive a password for free access to the full articles by registering at:

http://www.springerlink.com

If you do not have a subscription, you can still view the tables of contents of the volumes and the abstract of each article by going to the SpringerLink Homepage, clicking on "Browse by Online Libraries", then "Chemical Sciences", and finally choose Topics in Current Chemistry.

You will find information about the

- Editorial Board
- Aims and Scope
- Instructions for Authors
- Sample Contribution

at http://www.springeronline.com using the search function.

Preface

How can an enzyme that apparently does not utilize cofactors or covalent intermediates be one of the most proficient enzymes known? This is the mystery of orotidine monophosphate decarboxylase (ODCase). In this volume, experts in the field of enzyme catalysis describe their efforts to understand this puzzling enzyme.

The reaction catalyzed by ODCase is ostensibly quite simple—the decarboxylation of orotate ribose monophosphate (OMP) to produce uracil ribose monophosphate (see below). The problem with this reaction is that direct decarboxylation would lead to an anion whose lone pair of electrons is not aligned with any π-system that could lead to stabilization through delocalization. Various proposals have been put forth to overcome this apparent obstacle. These include selective stabilization of the transition state for direct decarboxylation by noncovalent interactions with ODCase, selective destabilization of the reactant by repulsive noncovalent interactions that are reduced or removed during direct decarboxylation, pre-protonation of the reactant on one of its carbonyl oxygens or alkene carbons such that decarboxylation would lead to a stabilized ylide, and concerted protonation-decarboxylation which would avoid the formation of a discrete uracil anion. The validity of these mechanistic proposals is analyzed from various viewpoints in this volume.

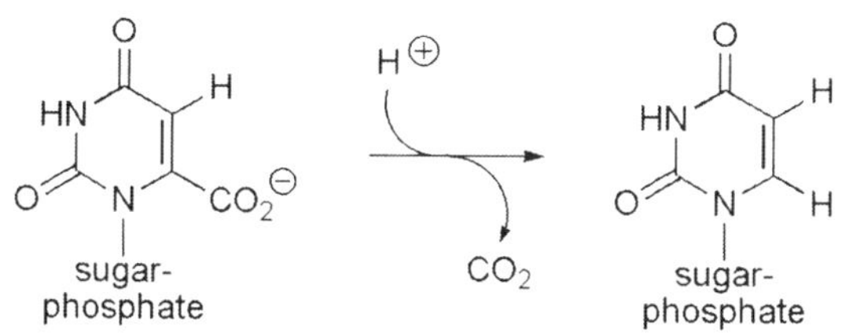

Houk and coworkers provide a survey of the known biochemical, structural, and computational studies on ODCase. In particular, they examine what recent theoretical studies have discovered about the origins of catalysis in ODCase, including the possibility that dynamic effects and/or iminium ion formation might actually be important contributors to ODCase's proficiency.

Wu and Pai provide a thorough examination of the molecular structure of ODCase, based on X-ray crystallographic studies of wild-type and mutant ODCases bound with various inhibitors. The implications of these structural studies are discussed, with an emphasis on the possibility that the conformational dynamics of the enzyme may be the key to catalysis.

Miller provides a synthesis of results from recent structural studies, computations, and biochemical experiments on mutant ODCases and truncated substrates. His detailed analysis of specific ODCase-substrate interactions is aimed at quantifying their importance and proposing roles for each in catalysis.

Smiley provides an analysis of the available structural and biochemical data, and based on these, proposes an unusual binding mode for OMP in ODCase. The implications of this binding mode for catalysis are discussed in detail.

Lundberg, Blomberg, and Siegbahn show how modern quantum chemistry has been applied to the ODCase problem. They describe results from quantum mechanical calculations on large models of OMP and the active site residues in ODCase that surround it, and provide a critical evaluation of many of the proposed mechanisms for catalysis.

Gao and coworkers describe how combined quantum mechanics/molecular mechanics computations have also been applied to the ODCase problem. In particular, he focuses on the importance of strain (in both the substrate and enzyme) to catalysis of direct decarboxylation.

Overall, this volume shows how many different mechanistic tools—biochemical kinetics, X-ray crystallography, state-of-the-art computations—have been brought to bear in parallel and occasionally in concert on the problem of understanding the origins of catalysis by ODCase. Still, it is clear that a consensus has not yet been reached as to the catalytic mechanism. In fact, while very few mechanistic proposals have been disproven, many new mechanistic proposals have arisen. For example, as pointed out by several of our authors, dynamic effects may prove to be essential for catalysis in this peculiar enzyme. We hope that the diverse chemistry discussed in this volume will inspire not only additional calculations and experiments, but also new viewpoints from which the mystery of ODCase's catalytic mechanism may be unraveled.

Davis and Piscataway, March 2004 Dean Tantillo and Jeehiun Lee

Contents

Top Curr Chem (2003) 238:1–22
DOI 10.1007/b94536

What Have Theory and Crystallography Revealed About the Mechanism of Catalysis by Orotidine Monophosphate Decarboxylase?

K. N. Houk[1] (✉) · Dean J. Tantillo[2] · Courtney Stanton[1] · Yunfeng Hu[1]

[1] Department of Chemistry and Biochemistry, University of California, Los Angeles, CA 90095–1569, USA
houk@chem.ucla.edu; cstanton@chem.ucla.edu; yunfeng@chem.ucla.edu

[2] Department of Chemistry and Chemical Biology, Baker Laboratory, Cornell University, Ithaca, NY 14853–1301, USA
Current address:
Department of Chemistry, University of California, One Shields Avenue, Davis, CA 95616, USA
dt64@cornell.edu

Abstract In 1995, Wolfenden and Radzicka showed that orotidine 5′-monophosphate decarboxylase (ODCase) was the most proficient enzyme. Since then, the mechanism of catalysis has been widely debated. The recent appearance of crystal structures for ODCase has led not to a definitive picture of catalysis as might be expected, but to even more conjecture concerning the mechanism. In addition, the many theoretical studies on ODCase have caused opinions to diverge, rather than converge, about the mechanism. This review summarizes the mechanistic, crystallographic, and computational evidence for the mechanism of ODCase, and offers a critical evaluation of the various mechanisms based upon this evidence.

Keywords Quantum mechanics · Density functional theory · ODCase · Uracil · Decarboxylation

Abbreviations

ODCase	Orotidine 5′-monophosphate decarboxylase
OMP	Orotidine 5′-monophosphate
UMP	Uridine 5′-monophosphate
BMP	Barbituric acid ribosyl 5′-monophosphate
6-azaUMP	6-azauridylate 5′-monophosphate
XMP	Xanthosine 5′-monophosphate
CMP	Cytidine 5′-monophosphate
CPMD	Car-Parrinello molecular dynamics

1 Introduction

The mechanism of the enzymatic decarboxylation of orotidine 5′-monophosphate (OMP) to uridine 5′-monophosphate (UMP) (see Fig. 1) is an intriguing problem for which many solutions have been offered. Even before 1995 when Wolfenden and Radzicka declared OMP decarboxylase (ODCase) to be the "most proficient enzyme" [1], several different mechanisms had been proposed. Since that time, other mechanisms have been advocated. Curiously, the appearance of crystal structures for various wild-type and mutant ODCases has led not to a definitive picture of catalysis, but to even more conjecture and controversy concerning the mechanism.

This article summarizes the mechanistic, crystallographic, and computational evidence for the mechanism of decarboxylation of OMP by the family of orotidine 5′-monophosphate decarboxylase enzymes, and offers a critical evaluation of the various mechanisms based upon this evidence.

Figure 2 summarizes the nine different types of mechanisms proposed for this reaction and the original proponents of each. There is some overlap,

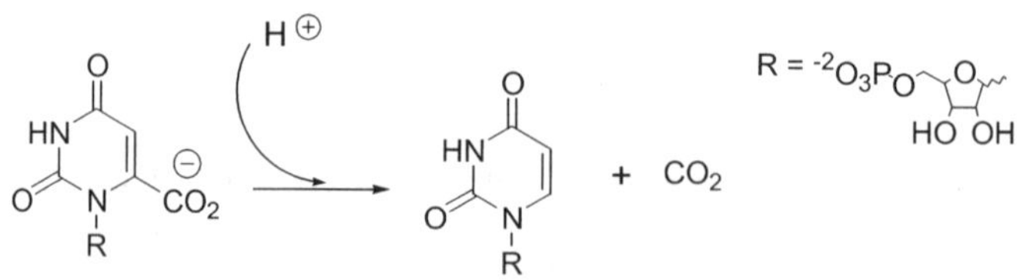

Fig. 1 Conversion of orotidine monophosphate to uridine monophosphate

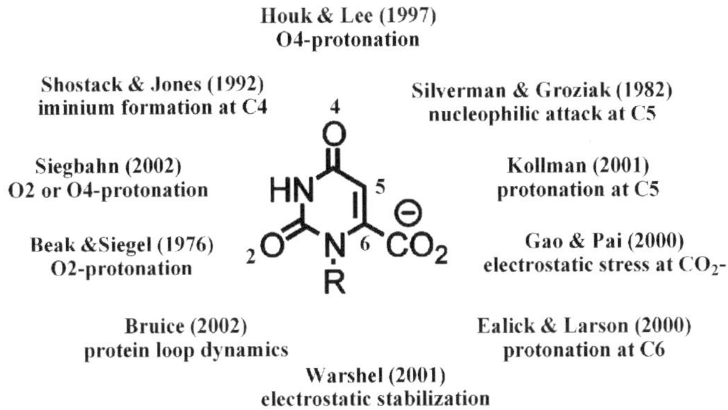

Fig. 2 Summary of mechanisms proposed for action of ODCase

and more than one researcher has backed one of the other mechanisms. Nevertheless, consensus has not been achieved.

Figure 3 compares the proficiencies ($k_{cat}/K_M/k_{un}$) of ODCase, several other enzyme decarboxylases [2], and some antibody decarboxylases [3]. The proficiencies of the decarboxylase enzymes, including a variety of amino acid decarboxylases, are nearly equal. Many decarboxylases employ iminium intermediates formed by reaction of an amino acid with a cofactor such as pyruvoyl or pyridoxal, or by reaction of a *β*-keto ester with an active-site lysine residue. These intermediates have been found to be so reactive that the

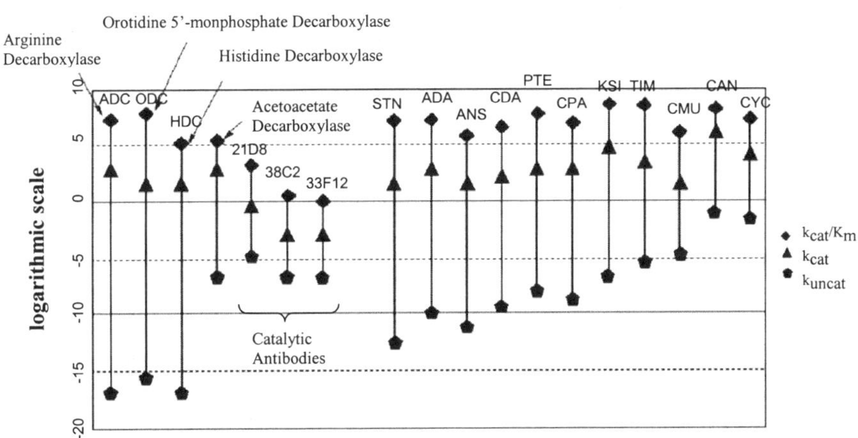

Fig. 3 A modified "Wolfenden plot" showing k_{cat}/K_M, k_{cat}, and k_{uncat} for a selection of enzymes and antibodies: Staphylococcal nuclease (*STN*), adenosine deaminase (*ADA*), cytidine deaminase (*CDA*), alanylalanine transpeptidase (*PTE*), carboxypeptidase A (*CPA*), ketosteroid isomerase (*KSI*), triose phosphate isomerase (*TIM*), chorismate mutase (*CMU*), carbonic anhydrase (*CAN*), cyclophilin (*CYC*)

rate of formation of the iminium intermediate becomes competitive with the rate of decarboxylation [4].

For ODCase, non-covalent mechanisms have often been proposed, as reflected in three of the mechanisms shown in Fig. 2. This is the crux of the attention showered on ODCase: how can this enzyme achieve its rate acceleration without the use of cofactors, metals, or acid-base catalysis? From Wolfenden's measurements of the uncatalyzed reaction of 1-methylorotic acid in water, he calculated the rate enhancement (k_{cat}/k_{un}) in the enzyme to be 1.4×10^{17}, corresponding to a reduction of $\Delta G^{\ddagger}$ of 24 kcal/mol [1]. He also reported the catalytic proficiency to be 2×10^{23}, meaning that the enzyme–transition state complex is an impressive 32 kcal/mol more stable than the free enzyme and transition state in water (i.e., the effective binding free energy of the transition state out of water is 32 kcal/mol) [1]! The experimental free energy of activation is 15 kcal/mol for this decarboxylation in ODCase.

2 Mechanistic Investigations of ODCase

The lack of evidence for the involvement of cofactors in ODCase catalyzed decarboxylation prompted Beak and Siegel to propose a novel mechanism, which was supported by experiments on model systems [5]. They showed that the decarboxylation of 1,3-dimethylorotic acid in sulfolane follows two different pH-dependent pathways. In neutral solution, decarboxylation is initiated by zwitterion formation. This led to the proposal of a zwitterion intermediate in the ODCase catalyzed reaction in which the O2 of the pyrimidine ring is protonated, promoting the release of CO_2.

Various covalent mechanisms have also been postulated. Silverman and Groziak used model systems to support the Michael addition of an active-site nucleophile to the C5 position of OMP followed by decarboxylation [6]. This mechanism was later discarded when kinetic isotope effect experiments indicated that there was no change in bond order between C5 and C6 during decarboxylation [7].

As noted above, many decarboxylases are known to exploit Schiff base formation in the active-site as a source of catalysis. Shostack and Jones explored this possibility in the case of ODCase [8]. They found that when the enzymatic reaction is performed in ^{18}O water, the product does not incorporate ^{18}O from bulk solvent. For this reason, a covalent iminium mechanism for ODCase was abandoned, in spite of its attractive similarities to other decarboxylase mechanisms.

Besides structural characterization described below, various other experimental techniques have been employed to probe the nature of the

ODCase active site. Wolfenden performed atomic absorption spectroscopy and initially concluded that two atoms of zinc were present per active ODCase monomer [9]. This finding was later withdrawn when atomic and x-ray absorption spectroscopy did not detect zinc in active ODCase samples [10].

Mutagenesis experiments were also used to determine the importance of active-site residues involved in binding and catalysis. Mutation of either Lys93 or Asp91 (yeast numbering) to alanine was shown to reduce activity by more than 10^5, and to drastically reduce substrate binding [11]. The substitution of alanine at Lys59 or Asp96 reduced the activity by a comparable amount, but the mutant retained some substrate affinity.

Likewise, a series of mutagenesis and substrate truncation experiments indicated the importance of phosphate binding to the complexation of substrate and to catalysis [11, 12]. Single mutations of several active-site residues that are known to interact with the phosphoryl group reduced the value of k_{cat}/K_m by factors of up to 7,300-fold. The value of k_{cat}/K_m was reduced by more than twelve orders of magnitude when the phosphoribosyl group of the OMP substrate was removed. The magnitude of this interaction has been used to argue in favor of substrate destabilization as a source of catalysis, whereby the favorable interaction of the phosphoryl group compensates energetically for the destabilization of the reactive portion of the substrate (discussed in detail in Sect. 4). More recent mutagenesis experiments indicate that favorable interactions between the enzyme and substrate phosphoryl group increase as the substrate approaches the transition state, reinforcing the importance of transition state stabilization in catalysis [13].

3 Crystallographic Studies

3.1 Initial Reports

To date, more than twenty X-ray crystal structures of ODCase have been reported, all by the groups of Ealick, Pai, Larsen, and Short and Wolfenden (see Table 1) [14–21]. The first six of these were reviewed previously by Houk, Lee, and coworkers [22] and Begley, Appleby, and Ealick [23]. As noted by these reviewers, the active-site structure of ODCase is extremely similar in all of the reported structures, boasting an Asp-Lys-Asp-Lys tetrad in the vicinity of the 6-position of the pyrimidine ring of bound inhibitors defined above, a group of uncharged hydrogen bond donors and acceptors that interact with the 2-, 3-, and 4-positions of the pyrimidine ring, and several hydrophobic residues that interact with the faces of the ring (Table 2). A representative structure is shown in Fig. 4.

Table 1 X-ray crystal structures of ODCase, listed in the order reported

Date	Ligand bound	Mutation(s)	Organism	Resolution (Å)	Group
2000	UMP	none	*B. subtilis*	2.4	Ealick [14]
2000	none	none	*S. cerevisiae*	2.1	Short & Wolfenden [15]
2000	BMP	none	*S. cerevisiae*	2.4	Short & Wolfenden [15]
2000	none	none	*M. thermoautotrophicum*	1.8	Pai [16]
2000	6-azaUMP	none	*M. thermoautotrophicum*	1.5	Pai [16]
2000	BMP	none	*E. coli*	2.5	Larsen [17]
2001	BMP	All Met to SeMet	*E. coli*	3	Larsen [18]
2002	6-azaUMP	Lys42Ala	*M. thermoautotrophicum*	1.5	Pai [19]
2002	6-azaUMP	Asp70Gly	*M. thermoautotrophicum*	1.5	Pai [19]
2002	UMP	Asp70Ala	*M. thermoautotrophicum*	1.5	Pai [19]
2002	6-azaUMP	Asp70Asn	*M. thermoautotrophicum*	1.5	Pai [19]
2002	UMP	Lys72Ala	*M. thermoautotrophicum*	1.5	Pai [19]
2002	OMP	Asp70Ala,Lys72Ala	*M. thermoautotrophicum*	1.5	Pai [19]
2002	6-azaUMP	Asp75BAsn	*M. thermoautotrophicum*	1.5	Pai [19]
2002	6-azaUMP	Ser127Ala	*M. thermoautotrophicum*	1.6	Pai [19]
2002	6-azaUMP	Gln185Ala	*M. thermoautotrophicum*	1.5	Pai [19]
2002	none	none	*E. coli*	2.5	Larsen [20]
2002	UMP	none	*M. thermoautotrophicum*	1.5	Pai [21]
2002	BMP	none	*M. thermoautotrophicum*	1.6	Pai [21]
2002	XMP	none	*M. thermoautotrophicum*	1.9	Pai [21]
2002	CMP	none	*M. thermoautotrophicum*	1.9	Pai [21]
2002	6-azaUMP	Arg203Ala,184–87 deleted	*M. thermoautotrophicum*	1.9	Pai [21]

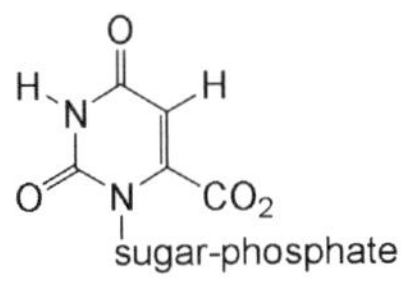

OMP

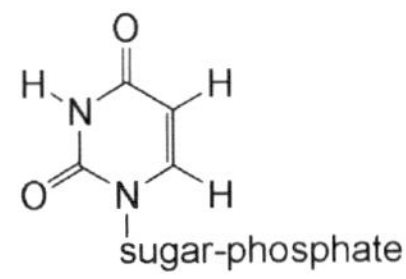

UMP

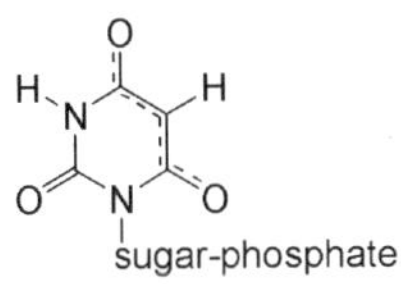

BMP

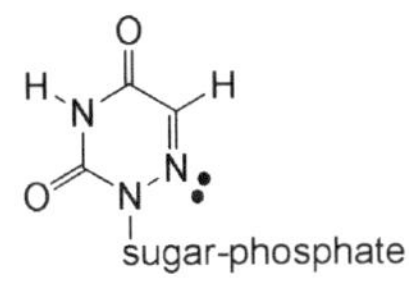

6-azaUMP

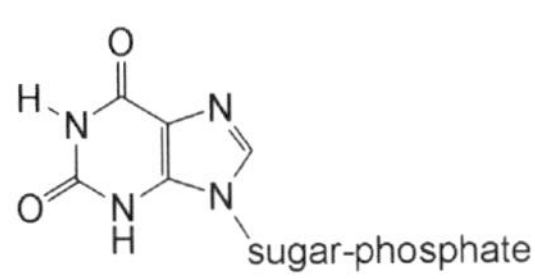

XMP

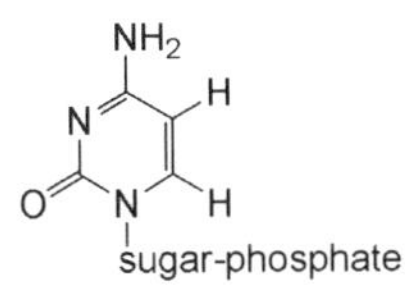

CMP

Although no structures of ODCase complexed with its natural substrate (OMP) were available at the time of these reports, several structures complexed to related compounds were described. These compounds (Chart 1) were UMP (the ultimate product of decarboxylation), BMP (a transition state analog in which a negatively charged oxygen atom replaces the carboxylate group of OMP), and 6-azaUMP (a transition state analog in which a nitrogen atom and its lone pair replace the C6-CO^- substructure of OMP and mimic the C6 anion thought to be produced upon OMP decarboxylation).

Table 2 Active-site residues in the pyrimidine-binding region

B. subtilis	*S. cerevisiae*	*M. thermoautotrophicum*	*E. coli*
Asp-Lys-Asp-Lys tetrad			
Lys33	Lys59	Lys42	Lys44
Asp60	Asp91	Asp70	Asp71
Lys62	Lys93	Lys72	Lys73
Asp65b	Asp96b	Asp75b	Asp76b
Uncharged hydrogen bond donors and acceptors			
Gln 194	Gln215	Gln185	Gln201
Thr123	Ser154	Ser127	Thr131
Hydrophobic residues			
Ile66b	Ile97b	Ile76b	Ile77b
Pro182	Pro202	Pro180	Pro189
Leu122	Leu153	Met126	Leu130
Val160	Ile183	Val155	Val167

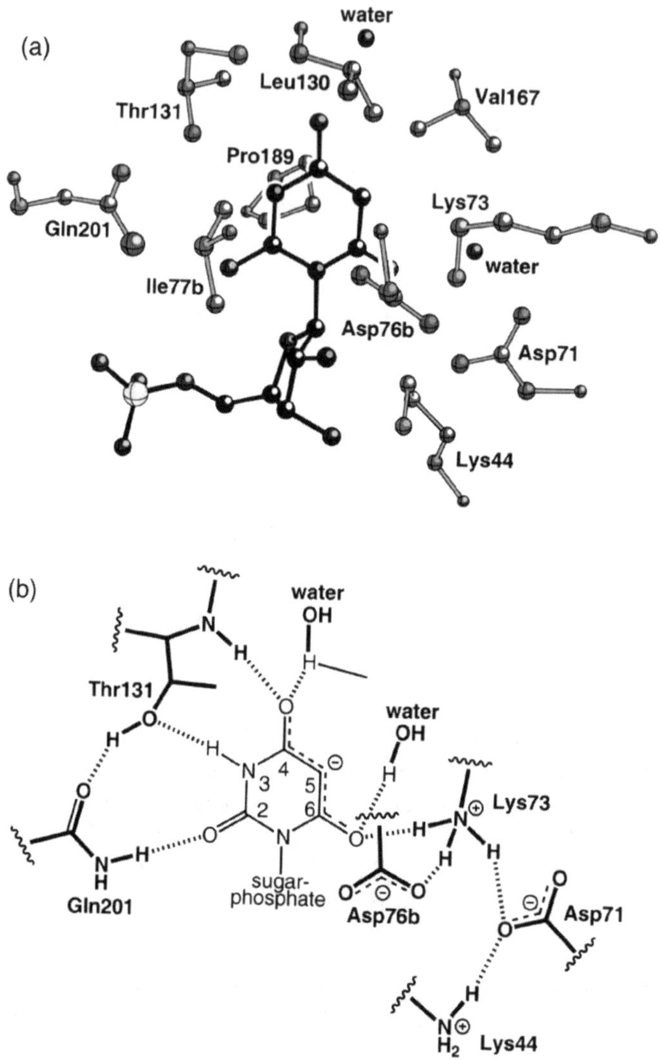

Fig. 4a,b ODCase active-site (pyrimidine-binding region) with BMP bound from *E. coli*. **a** Ball-and-stick drawing of polar and hydrophobic residues based on the crystallographically determined coordinates. Bonds in BMP are darker than those in the active-site residues. **b** Line drawing of polar active-site residues (*bold*) and their interactions with bound BMP

These ODCase-inhibitor structures provided some hints about the binding modes of species on the putative reaction coordinate for OMP decarboxylation. Overall, these structures showed that groups at the 6-position of a bound pyrimidine ring are in close proximity to the Asp-Lys-Asp-Lys tetrad, and when these groups contain lone pairs (as in BMP and 6-azaUMP), they hydrogen bond to a protonated Lys residue from the tetrad (Lys73 in Fig. 4, *E. coli* numbering). Such an arrangement might also occur for the carboxylate group of OMP, although this group is considerably larger than the corresponding functionality in BMP and 6-azaUMP. Uncertainty about the exact position of the carboxylate group in bound OMP has led to a debate over whether this carboxylate group would be stabilized upon complexation (through a hydrogen bond with the Lys or perhaps even a nearby Asp if it were protonated) or actually destabilized due to electrostatic repulsion with a deprotonated Asp. These two scenarios are discussed further in Sect. 4.

These initial structures also indicated that there is no strongly acidic residue in the vicinity of the carbonyl groups at the 2- or 4-position of the pyrimidine rings of any of the bound inhibitors (see Fig. 4). This observation cast doubt on the validity of earlier mechanistic proposals by Beak and Siegel and by Lee and Houk that invoked protonation of one or the other of these carbonyls in order to stabilize the charge that accumulates on OMP's pyrimidine ring during decarboxylation. Still, the possibility of a conformational change that brings an acidic residue closer to one of these carbonyls (for example, Lys73 moving near O4) or a proton transfer mediated by an active site water molecule (several of which are observed in these structures; see Fig. 4) have not been ruled out. Computational studies that address some of these issues are described in Sects. 4.1 and Sect. 4.5.

Additionally, comparison of unliganded ODCase structures with those of ODCase-inhibitor complexes indicated that ODCase likely undergoes a large conformational change upon complexation. This conformational change causes the active site to become sequestered from bulk solvent while several active-site residues move to assume positions suited for tight ligand binding. In particular, these movements involve residues that bind to the phosphate group and residues that participate in hydrophobic contacts with the pyrimidine ring. Initial computational studies on this conformational change and its implications for catalysis are discussed in Sect. 4.4.

3.2 Recent Studies

A large number of additional structures have been reported since these original six structures were described in 2000 (Table 1) [18, 19, 20, 21]. These more recent structures fall into three groups: (a) structures of ligand-free ODCase, (b) structures of wild-type ODCase bound to inhibitors, and (c) structures of ODCase mutants.

3.2.1
Ligand-Free ODCase

In 2002, the third structure of ligand-free ODCase was reported (Table 1) [20]. Again, significant conformational changes of some loops were observed when this structure was compared to the previously reported ODCase-BMP structure from the same organism, indicating that a large conformational change upon substrate binding is a general feature of ODCases.

3.2.2
Inhibitor Complexes

Also in 2002, Wu and Pai reported new structures of ODCase bound to UMP and BMP and also to the inhibitors XMP and CMP (Chart I) [21]. The structures of ODCase with UMP and BMP bound were extremely similar to those reported previously for ODCase from other organisms. The complexes of ODCase with XMP and CMP, however, differed considerably from those with UMP, BMP, and 6-azaUMP (for example, compare Figs. 4 and 5). Although many contacts between ODCase and the sugar-phosphate groups of XMP and CMP mirror those observed with UMP, BMP, and 6-azaUMP, the bases of XMP and CMP actually protrude from the binding region utilized by the other inhibitors and are consequently exposed to solvent. This change in binding mode is presumably a result of the fact that the xanthine and cyto-

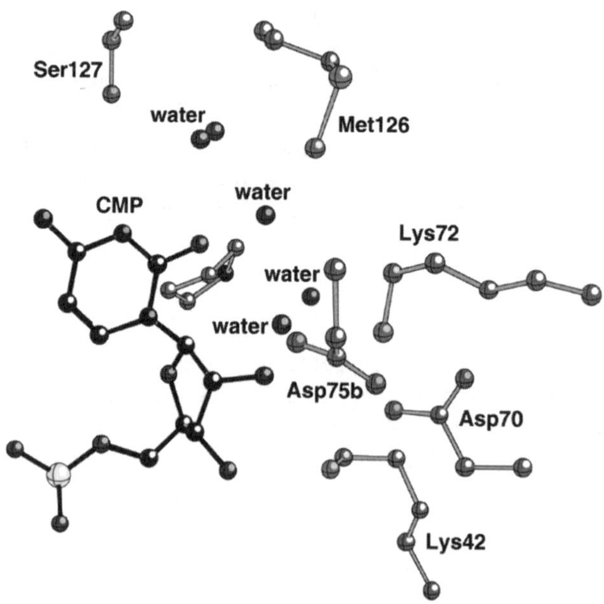

Fig. 5 ODCase active site (*M. thermoautotrophicum*) with CMP bound

sine rings in XMP and CMP, respectively, differ significantly from the pyrimidine rings in UMP, BMP, and 6-azaUMP in terms of size and hydrogen bonding pattern. This altered binding mode is also consistent with the fact that 6-carboxy-CMP is not a competent decarboxylation substrate for ODCase [24]. The implications of this binding mode for the mechanism by which ODCase catalyzes the decarboxylation of OMP—which is also considerably larger than UMP, BMP, and 6-azaUMP—are, as of yet, unclear. Wu and Pai do suggest, however, that this newly observed binding mode may reflect a metastable binding orientation for OMP that occurs between initial binding of the phosphate group and binding of the full phosphate–sugar–orotate structure.

3.2.3
ODCase Mutants

Pai and coworkers also reported structural studies of several *M. thermoautotrophicum* ODCase mutants (see Table 1) [19, 21]. These structural studies focused on how binding is affected when residues in the phosphate binding region, the recognition site for the 2- through 4-positions of the pyrimidine ring of bound inhibitors, and the Asp-Lys-Asp-Lys tetrad (see Fig. 4 and Table 2) are altered.

A structural study of a mutant ODCase in which a key phosphate-binding Arg was truncated to Ala and a portion of a loop containing other phosphate binding residues was deleted was reported by Wu and Pai in 2002 [21]. Surprisingly, 6-azaUMP was observed to bind to this mutant in the same orientation as in its complex with wild-type ODCase, despite the disruption of several ODCase-phosphate interactions. Nonetheless, the binding affinity of 6-azaUMP for this mutant was decreased relative to that for the native structure and the crystals formed with the mutant ODCase were considerably less stable than those formed with the native enzyme, consistent with other evidence indicating that phosphate binding is essential for effective catalysis (see Sect. 2).

Structures of 6-azaUMP-ODCase complexes in which either Ser127 or Gln185 (analogous to Thr131 and Gln201 in the *E. coli* structure shown in Fig. 4; see also Table 2) were mutated to Ala suggested that the side-chain of Gln185 is not essential for effective recognition of the 2- through 4-positions of the pyrimidine ring, while that of Ser127 is [19]. When Gln185 was mutated to Ala, a string of well-ordered water molecules was observed to stand in for the amide group found in the native structure, allowing 6-azaUMP to bind in the same orientation as it had before mutation. In contrast, when Ser127 was mutated to Ala, a roughly equal mixture of two 6-azaUMP complexes was observed—half had 6-azaUMP occupying the same position it had in the wild-type structure, and half had its azauracil ring flipped by approximately 180° around the base-sugar N–C bond.

Several structures of ODCase in which components of the Asp-Lys-Asp-Lys tetrad were mutated were also reported by Pai and coworkers in 2002 [19]. When Asp70 (analogous to Asp71 in the *E. coli* structure shown in Fig. 4; see also Table 2) was mutated to Ala and the crystallization medium contained OMP, a complex of UMP—the product of OMP decarboxylation—was observed. In this complex, a chloride ion from the medium assumed the bridging position between the two Lys residues of the tetrad that was occupied by the carboxylate group of Asp70 in the wild-type structure [19]. It is tempting to use the fact that this mutant could still convert OMP to UMP to argue against a ground state destabilization mechanism involving electrostatic repulsion between Asp70 and the carboxylate of OMP. However, the implications of this result for the catalytic mechanism of wild-type ODCase in its natural environment are unclear, given the presence of the chloride ion and the relatively long time-scale of the crystallization experiments. Mutation of Asp70 to Gly and crystallization with 6-azaUMP lead to a mixture of structures displaying several conformers for the Lys72 side-chain and the azauracil ring (that observed in the wild-type structure and the rotamer with the ring flipped by approximately 180°) [19]. Finally, mutation of Asp70 to an isosteric but neutral Asn residue resulted in a 6-azaUMP complex in which the azauracil ring was flipped by approximately 180° compared to its conformation in the wild-type structure [19]. The nearly exclusive population of this alternative conformer is perhaps driven by the formation of a hydrogen bond between the oxygen at the 2-position of the azauracil ring and the positively charged Lys72, which should be a stronger hydrogen bond donor here than it is in the presence of the negatively charged Asp70 found in wild-type ODCase. When the other Asp of the tetrad, Asp75B, is mutated to Asn, a 6-azaUMP complex is observed in which the amide side-chain of the Asn is rotated away from the tetrad and replaced again by a chloride ion from the crystallization solution [19].

The two Lys residues of the Asp-Lys-Asp-Lys tetrad were also mutated, each to Ala [19]. In the structure of the Lys42Ala mutant complexed to 6-azaUMP, the position normally occupied by the ammonium group of the Lys was filled by a water molecule, and 6-azaUMP was bound in the same orientation as in the wild-type structure. Crystallization of the Lys72Ala mutant in the presence of OMP led again to a complex with UMP, indicating that this mutant also does not completely lose its ability to promote decarboxylation. Again, the implications of this result are unclear given the length of the crystallization experiments and the composition of the crystallization medium. In this mutant, density corresponding to a water molecule appears at the position normally occupied by the ammonium group of Lys72, analogous to the situation observed for the Lys42Ala mutant.

Simultaneous mutation of Lys72 and Asp70 both to Ala resulted in the first reported ODCase-OMP complex [19]. In this complex, the carboxylate of OMP, along with a water molecule, occupies space normally occupied by the side-chains of Lys72 and Asp70 and the pyrimidine ring is bound in the usual orientation. It had been noted previously that the carboxylate group of OMP could not fit comfortably in the unmutated binding site without some conformational change. Removal of the Lys72 and Asp70 side-chains apparently relieves this congestion but also destroys the ability of the enzyme to decarboxylate OMP. Although Pai and coworkers use this information to argue for a ground state destabilization mechanism, a better understanding of the actual binding mode of OMP to wild-type ODCase is required before firm conclusions about the ODCase mechanism can be reached.

4 Theoretical Studies of ODCase

In general, decarboxylase enzymes stabilize carbanion intermediates formed upon decarboxylation via delocalization of their anionic electron pairs into attached π systems [25, 26, 27]. However, direct decarboxylation of orotidine 5′-monophosphate (OMP) would form a non-conjugated carbanion (at C6) that is not stabilized in such a manner. The absence of any metal ions or cofactors in ODCase, such as the pyruvoyl substituent or pyridoxal 5′-phosphate cofactor, which could aid charge delocalization, is thus even more surprising. The following sections describe various computational studies aimed at elucidating the origins of catalysis by the seemingly unique ODCase.

4.1 Oxygen Protonation

Beak and Siegel proposed the formation of a zwitterionic intermediate upon decarboxylation, which could provide rate acceleration through "dipole-stabilization" (see Fig. 6, structure 1) [7]. A related mechanism was put forth 20 years later when Lee and Houk proposed protonation at O4 instead of O2 (Fig. 6, structure 2) [27]. Quantum mechanical calculations on model systems (R=H in Fig. 6) were used by Lee and Houk to calculate and compare reaction energetics for these two pathways. It was shown that the activation enthalpy of the parent reaction was reduced from 44 kcal/mol in the gas phase to 22 kcal/mol with O2 protonation. The barrier was further reduced upon O4 protonation, to a value of 15.5 kcal/mol, indicating that O4 protonation is a viable strategy for catalysis. The preference for O4 protonation can be explained, at least in part, by examining the reasonable resonance structures of the decarboxylated intermediates (see Fig. 6). While every resonance structure for the O2 protonated intermediate involves charge separation, O4 protonation yields a stabilized neutral carbene.

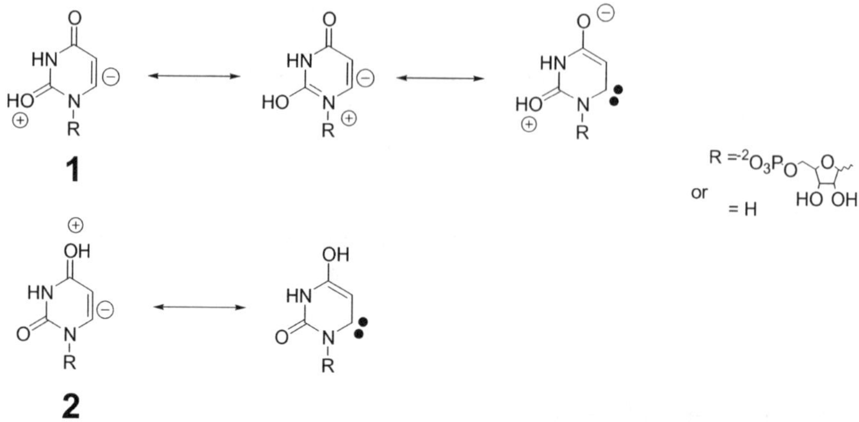

Fig. 6 The *top* resonance structures (*1*) represent the O2-protonated zwitterions. The *bottom* resonance structures (*2*) represent the O4-protonated zwitterion and stabilized neutral carbene. The R group was changed to hydrogen for the calculations described in the text

Additional density functional calculations on orotate derivatives by Singleton et al. indicated that the major factor favoring decarboxylation via O4 protonation is likely an inherent preference for O4 protonation in uracil derivatives, rather than strong selective stabilization of the O4-protonated decarboxylated product and the transition state for its formation [28]. The authors also compared experimental (in solution without enzyme present) and theoretical ^{13}C kinetic isotope effects, finding that those computed for the O4 protonation pathway matched best the experimentally determined values. It was also noted, though, that differences between isotope effects measured for the carboxylate carbon in the uncatalyzed [28] and enzyme-catalyzed [29] decarboxylations may indicate that the mechanisms in these two environments differ considerably. More recent ^{15}N isotope effect calculations performed by Phillips and Lee [30], however, indicate that protonation of either O4 or O2 is consistent with the reported experimental ^{15}N isotope effects for the ODCase-catalyzed reaction [31].

After the initial crystal structures of ODCase were reported, a modified O4 protonation mechanism was proposed [22]. In the original Lee-Houk proposal, the active-site residue Lys93 (yeast numbering, analogous to Lys72 in *E. coli*; Fig. 4 and Table 2), which was experimentally shown to be catalytically important, was assumed to be the species that protonated O4 [27, 32]. However, in the reported crystal structures this lysine does not reside near O4. It was therefore proposed that proton transfer from Lys93 to O4 might be mediated by an active site water molecule (see Fig. 7). The viability of this scenario has not yet been established, however.

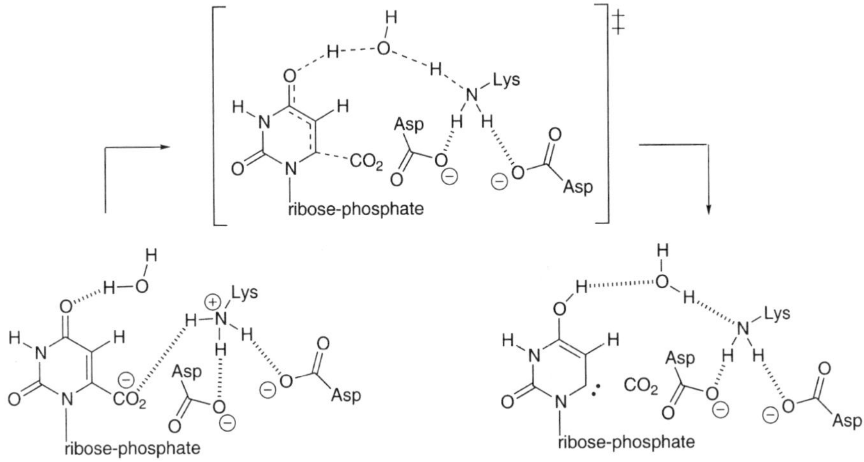

Fig. 7 Proposed modification of the Lee-Houk O4-protonation mechanism

4.2
Ground State Destabilization or Transition State Stabilization?

Gao, Pai, and coworkers conducted a hybrid QM/MM study which combined semiempirical quantum mechanical AM1 calculations on the substrate with CHARMM22 force field calculations on the surrounding protein environment [16]. These calculations were able to reproduce the experimental activation free energies for both the uncatalyzed and catalyzed decarboxylation reactions. In their analysis, OMP was considered to be composed of two parts: the reactive part (S_R), corresponding to the pyrimidine ring, and the unreactive part (S_b), corresponding to the ribose and phosphate groups. They proposed that the strong binding of S_b with ODCase leaves S_R in an energeticaly unfavorable position—close to one negatively charged Asp of the Asp-Lys-Asp-Lys tetrad; in other words, the pyrimidine portion of the ground state is destabilized more than that of the transition state upon binding to ODCase, thus lowering the activation barrier. Such a mechanism has been called the Circe effect [16, 33]. Other groups have also considered similar mechanisms [14, 17, 34]. Recently Gao proposed that the enzyme conformational change that occurs upon binding (see Sect. 3) may introduce the Circe effect in a different way [35]. He hypothesized that when it binds the substrate, the enzyme adopts an energetically unfavorable conformation. This stress is then released while the reaction proceeds, resulting in rate acceleration.

Warshel and coworkers used ab initio quantum mechanical calculations at the HF/6–31G* and BLYP/AUG-cc-pVDZ levels to perform geometry optimizations and obtain single point energy values, respectively, for the decarboxylation of a model system made up of NH_4^+ and orotate [36, 37]. Empirical valence bond simulations (EVB) and Langevin dipole dynamics calculations were also used to conduct free energy perturbation calculations on a model of the whole OMP-ODCase system constructed from the crystal structure of the 6-azaUMP-ODCase complex [16] by substituting 6-azaUMP with OMP. These authors concluded that ODCase catalysis results from selective transition state stabilization rather than ground state destabilization. They note that in the Circe effect mechanism, approximately 32 kcal/mol of binding energy for S_b is needed to compensate for the destabilization of the orotate group, while experiments show that phosphate and ribose binding in fact contribute much less [12]. Therefore, they suggest instead that the protein active-site is preorganized to complement the transition state so that a large reorganization energy penalty need not be paid as decarboxylation proceeds.

In addition, although the close proximity of two aspartic acid residues to the likely position of the substrate carboxylate group observed in the reported crystal structures of ODCase (see Fig. 4 and Sect. 3) originally led to the hypothesis that electrostatic repulsion could lead to ground-state destabilization [16, 17], we feel that the various options available to enzymes to avoid such unfavorable interactions (such as carboxylate protonation), and the observed favorable interaction of the binding site with the negatively charged inhibitor BMP (see Fig. 4) render this mechanism unlikely [22].

4.3
C5 Protonation

Kollman and coworkers applied a variety of computational methods to this mechanistic problem—including quantum mechanics on small model systems, molecular dynamics simulations with the AMBER force field on the whole ODCase-substrate system, and MM-PBSA free energy calculations on ODCase with bound OMP [38]. Based on their results, they proposed a decarboxylation mechanism for ODCase that involves C5 protonation. Their calculations at the MP2/6–31+G*//HF/6–31+G* level showed that C5 has a greater intrinsic proton affinity than C6, O2, and even O4. This, coupled with the fact that Lys72 (*M. thermoautotrophicum* numbering; see Table 2) is near C5 and C6 in the inhibitor-bound crystal structures, prompted the authors to embrace a C5 protonation mechanism. However, the authors themselves acknowledged the uncertainties of their calculations because of approximations employed in representing the enzyme active site.

4.4
Loop Dynamics

Hur and Bruice carried out classical molecular dynamics on the complex of ODCase with OMP before decarboxylation, as well as the putative intermediate (the C6 anion) formed by decarboxylation [39]. Based on these calculations, it was proposed that loop movement in ODCase may play a key role in catalysis; a stable β hairpin structure appeared to form during decarboxylation that was not present before decarboxylation. In addition, the structure of OMP in aqueous solution was also simulated, and the similarity of the conformations of OMP in water and in the ODCase active site suggested that OMP is not bound in a particularly strained fashion, further arguing against a Circe effect.

4.5
Large Quantum Mechanical Models

Lundberg, Blomberg, and Siegbahn used density functional calculations to investigate several model systems composed of orotate and various key active site residues [40]. Based on these calculations, they too suggested that electrostatic stress and geometrical strain do not play significant roles in catalysis. In addition, they suggested that an O2 protonation mechanism might actually be possible—it is the most energetically favorable mechanism that they studied—although their calculated barrier for this mechanism is 25 kcal/mol, which is still ~10 kcal/mol higher than the experimentally determined barrier.

4.6
Car-Parrinello Molecular Dynamics

Static quantum mechanical calculations have respectable energetic accuracy but are limited to a single geometry per calculation. On the other hand, classical molecular dynamics calculations are able to take into account dynamic effects which might be critical to enzyme catalysis and are difficult to deal with by conventional quantum mechanics; these calculations employ empirical force fields, however. The Car-Parrinello molecular dynamics method (CPMD) [41, 42], though, has successfully incorporated the accuracy of quantum mechanical calculations into dynamics simulations at a reasonable cost. CPMD 3.5 [43] has been employed by our group to investigate various model systems related to the ODCase mechanism [44].

A molecular dynamics simulation on the ground state of orotate in a model enzyme pocket was performed. The simplest system studied included methyl orotate and models of Lys72, Asp70 (*M. thermoautotrophicum* numbering; see Chart I), and one crystallographic water as shown in Fig. 8. In

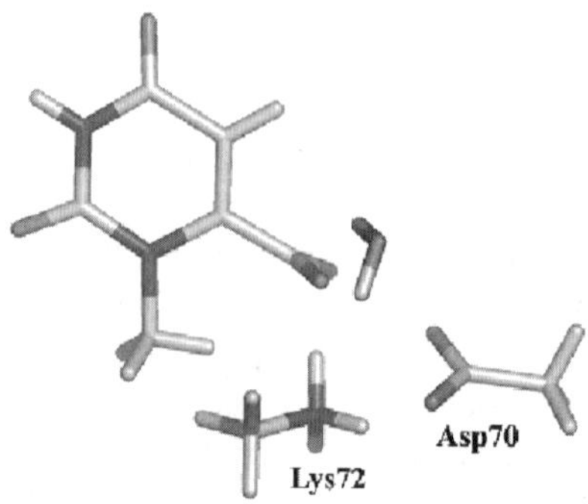

Fig. 8 Model system for a 20,000-step constrained molecular dynamics

constructing this model, the carboxylate group was positioned based on quantum mechanical optimizations and the location of inhibitors in the reported ODCase crystal structures (see Sect. 3). In this model, neither the water molecule nor Lys72 seem to make close contacts with the orotate C6. The CPMD simulations indicated that the aspartate is protonated by the lysine at a very early stage (see Fig. 9a and b), corresponding to a large drop in ener-

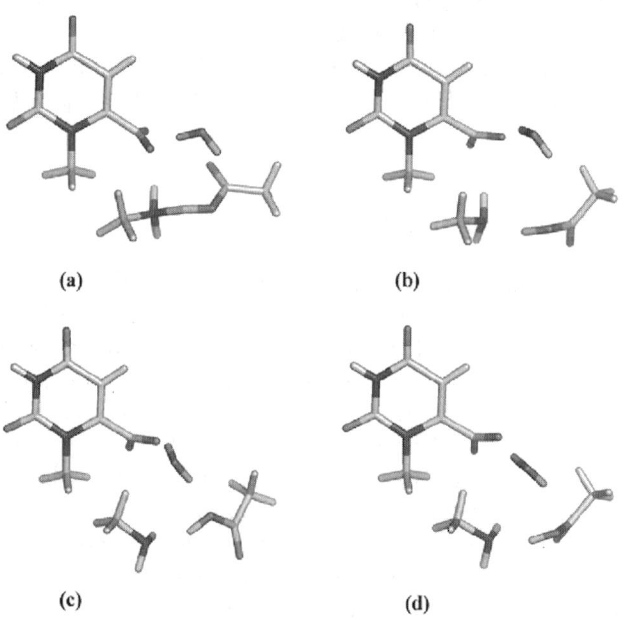

Fig. 9a–d Four snapshots: **a** salt-bridge is formed between Lys72 and Asp70; **b** proton is fully transferred to aspartate; **c** water bridges orotate carboxylate and aspartate while aspartate and lysine also form a hydrogen bond; **d** water forms a hydrogen bond with the orotate carboxylate, while aspartate and lysine form a hydrogen bond

gy. The aspartate then switches between two alternative orientations with (Fig. 9c) or without (Fig. 9d) a hydrogen bond to the water molecule. The ground state destabilization mechanism seems inconsistent with this delicate hydrogen bonding network formed between orotate, lysine, aspartate, and water. A second CPMD simulation was also conducted with a neutral lysine, giving similar results.

4.7 Iminium Ion Formation

Iminium ion formation often precedes decarboxylation and is essential for catalysis in many enzymes. As noted in the introduction, Shostack and Jones proposed a Schiff base mechanism for ODCase that involves iminium formation with an active-site lysine [8]. However, experiments have shown that no exchange between substrate and bulk solvent occurred when the enzymatic reaction was run in $H_2{}^{18}O$ [10].

Nonetheless, the importance of Lys93 in ODCase catalysis prompted our group to revisit theoretically the possibility of iminium formation as a potential mechanism. Model system calculations show that the reaction free energy decreases quite dramatically with Schiff base formation at C4 (Table 3), from 40.6 kcal/mol in the parent reaction to 21.2 kcal/mol in the gas phase, establishing the feasibility of such a mechanism [45]. Thus, it seems plausible that Lys93 (yeast numbering; see Table 2) might covalently attach to the substrate, resulting in a lowering of the barrier for decarboxylation. To determine if this is structurally feasible, a model was constructed in which the substrate is covalently bound to Lys93 in the open form of the enzyme (see Fig. 10), and such an arrangement does seem possible. Molecular dynamics simulations are currently underway to further explore iminium formation at this site as a possible mechanism [45].

It should be pointed out that the assumption that Schiff base formation should inevitably lead to exchange of solvent [^{18}O] during the course of an enzymatic reaction has been questioned. For many years, solvent [^{18}O] exchange experiments were used to distinguish between type I aldolases, which operate via formation of a Schiff base intermediate that links the sub-

Table 3 Reaction energetics for model systems of possible ODCase mechanisms. All calculations done at B3LYP/6–31+g(d,p). Solvent calculations use the CPCM implicit solvent model

	ΔH_{gas} (kcal/mol)	ΔG_{gas} (kcal/mol)	ΔG_{H2O} (kcal/mol)
(1)	50.5	40.6	44.6
(2)	30.1	21.2	35.4
(3)	22.0	13.6	30.5

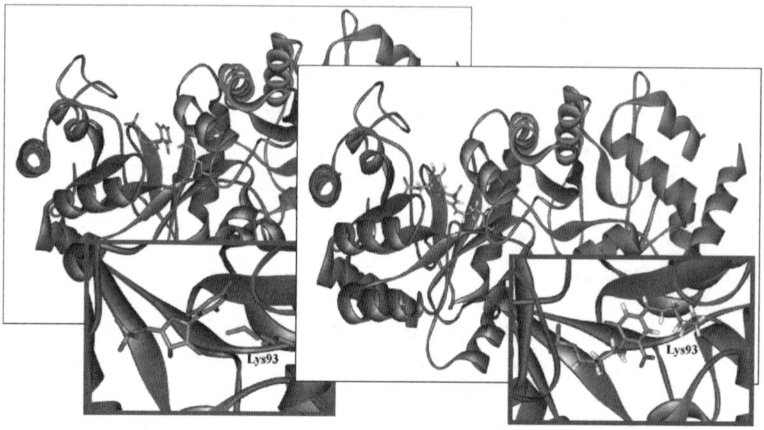

Fig. 10 *Left,* open form of ODCase dimer (1DQW) with superimposed BMP from closed form (1DQX, not shown). *Right,* open form of ODCase dimer (1DQW) with hypothetical model of substrate covalently bound to Lys93 (yeast numbering).

strate and an active site lysine, and type II aldolases, which employ Lewis acidic metal ions in catalysis. If the substrate exchanged [^{18}O] with the solvent, the enzyme was classified as type I, while if it did not exchange it was classified as a type II aldolase [46]. After the determination of several crystal structures for type I fructose bisphosphate aldolase, however, Littlechild and Watson proposed a Schiff base mechanism that they contend does not necessitate exchange with bulk solvent [47].

Beyond discussions of aldolases, there is a void in the literature concerning this issue. This is surprising considering that several Schiff base forming enzymes have been shown to catalyze their reactions with no solvent exchange in deuterated water. The pyridoxal 5′-phosphate-dependent enzyme lysine 2,3-aminomutase catalyzes the transfer of an amino group as shown in Fig. 11. Experimental evidence indicates that the proton abstracted from the substrate β-carbon is used to protonate the product α-carbon in a stepwise mechanism involving a Schiff base intermediate [48]. This would seem to indicate that a closed active-site may function to sequester the reacting species from the solvent medium. A similar situation may occur in ODCase.

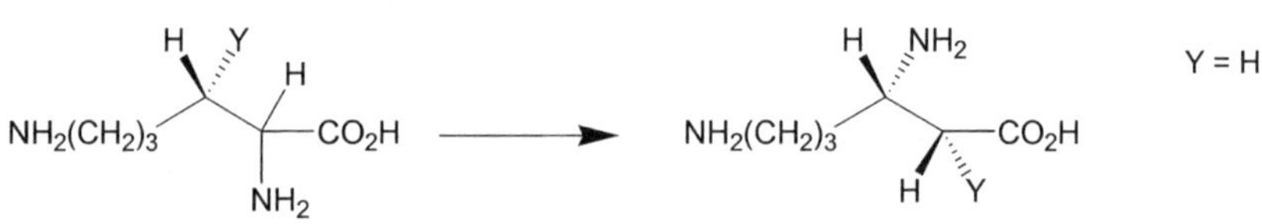

Fig. 11 Reaction catalyzed by lysine 2,3-aminomutase. See [48]

5 Conclusion

Despite intense experimental and theoretical efforts, the mystery surrounding the proficiency of ODCase has not yet been solved. This summary of the experimental facts and their implications, analysis of the structural data available from various crystal structures, description of the various computational studies performed on this problem, and discussion of the important mechanisms that have been proposed, shows the exciting challenges that remain in understanding the action of one of nature's most proficient enzymes.

References

1. Radzicka A, Wolfenden R (1995) Science 267:90
2. Snider MJ, Wolfenden R (2000) J Am Chem Soc 122:11507
3. Wolfenden R, Snider MJ (2001) Acc Chem Res 34:938
4. Zhou X, Jin X, Medhekar R, Chen X, Dieckmann T, Toney MD (2001) Biochemistry 40:1367
5. Beak P, Siegel B (1976) J Am Chem Soc 98:3601
6. Silverman RB, Groziak MP (1982) J Am Chem Soc 104:6434
7. Acheson SA, Bell JB, Jones ME, Wolfenden R (1990) Biochemistry 29:3198
8. Shostack K, Jones ME (1992) Biochemistry 31:12155
9. Miller BG, Traut TW, Wolfenden R (1998) J Am Chem Soc 120:2666
10. Cui W, DeWitt JG, Miller SM, Wu M (1999) Biochem Biophys Res Commun 259:133
11. Miller BG, Snider MJ, Short SA, Wolfenden R (2001) J Biol Chem 276:15174
12. Miller BG, Snider MJ, Short SA, Wolfenden R (2000) Biochemistry 39:8113
13. Miller BG, Butterfoss GL, Short SA, Wolfenden R (2001) Biochemistry 40:6227
14. Appleby TC, Kinsland C, Begley TP, Ealick SE (2000) Proc Nat Acad Sci USA 97:2005
15. Miller BG, Hassell AM, Wolfenden R, Milburn MV, Short SA (2000) Proc Nat Acad Sci USA 97:2011
16. Wu N, Mo Y, Gao J, Pai EF (2000) Proc Nat Acad Sci USA 97:2017
17. Harris P, Poulsen J-CN, Jensen KF, Larsen S (2000) Biochemistry 39:4217
18. Poulsen J-CN, Harris P, Jensen KF, Larsen S (2001) Acta Cryst D 57:1251
19. Wu N, Gillon W, Pai EF (2002) Biochemistry 41:4002
20. Harris P, Poulsen J-CN, Jensen KF, Larsen S (2002) J Mol Biol 318:1019
21. Wu N, Pai EF (2002) J Biol Chem 277:28080
22. Houk KN, Lee JK, Tantillo DJ, Bahmanyar S, Hietbrink BN (2001) ChemBioChem 2:113
23. Begley TP, Appleby TC, Ealick SE (2000) Curr Op Struct Biol 10:711
24. Smiley JA, Saleh L (1999) Bioorg Chem 27:297
25. Bruice TC, Benkovic S (1966) In: Bioorganic mechanisms, vol 2. Benjamin, New York, pp 188–194
26. Bender ML (1971) In: Mechanisms of homogeneous catalysis from protons to proteins. Wiley, New York, p 165 and p 586
27. Lee JK, Houk KN (1997) Science 276:942

28. Singleton DA, Merrigan SR, Kim BJ, Beak P, Phillips LM, Lee JK (2000) J Am Chem Soc 122:3296
29. Ehrlich JI, Hwang CC, Cook PF, Blanchard JS (1999) J Am Chem Soc 117:9357
30. Phillips LM, Lee JK (2001) J Am Chem Soc 123:12067
31. Rishavy MA, Cleland WW (2000) Biochemistry 39:4569
32. Smiley JA, Jones ME (1992) Biochemistry 31:12162
33. Jencks WP (1987) Catalysis in chemistry and enzymology. Dover, New York
34. Wolfenden R, Milburn MV, Short SA (2000) Proc Nat Acad Sci USA 97:2011
35. Gao J (2002) International society of quantum biology and pharmocology, President's conference, Snowbird, Utah 2002
36. Warshel A, Strajbl M, Villà J, Florián J (2000) Biochemistry 39:14728
37. Warshel A, Florián J, Strajbl M, Villà J (2001) ChemBioChem 2:109
38. Lee TS, Chong LT, Chodera JD, Kollman PA (2001) J Am Chem Soc 123:12837
39. Hur S, Bruice TC (2002) Proc Nat Acad Sci USA 99:9668
40. Lundberg M, Blomberg MRA, Siegbahn PE (2002) M J Mol Model 8:119
41. Car R, Parrinello M (1985) Phy Rev Lett 55:2471
42. Carloni P, Rothlisberger U, Parrinello M (2002) Acc Chem Res 35:455
43. Hutter J et al (2001) CPMD 3.5 (Copyright IBM Zurich Research Laboratory and MPI für Festkörperforschung 1995–2001)
44. Hu Y, Mundy C, Houk KN (unpublished results)
45. Houk KN, Stanton CL (unpublished results)
46. Heron EJ, Caprioli RM (1975) Biochim Biophys Acta 403:563
47. Littlechild JA, Watson HC (1993) Trends Biochem Sci 18:36
48. Aberhart DJ, Gould SJ, Lin HJ, Thiruvengadam TK, Weiller BH (1983) J Am Chem Soc 105:5461

Top Curr Chem (2004) 238:23–42
DOI 10.1007/b94537

Crystallographic Studies of Native and Mutant Orotidine 5′-Phosphate Decarboxylases

Ning Wu[1,2] (✉) · Emil F. Pai[2,3] (✉)

[1] Department of Biochemistry, University of Toronto, 1 King's College Circle, Toronto, Ontario, M5S 1A8, Canada
ning@hera.med.utoronto.ca

[2] Division of Molecular & Structural Biology, Ontario Cancer Institute/ University Health Network, 610 University Avenue, Toronto,Ontario, M5G 2M9, Canada
pai@hera.med.utoronto.ca; ning@hera.med.utoronto.ca

[3] Departments of Biochemistry, Medical Biophysics and Molecular & Medical Genetics, University of Toronto, 1 King's College Circle, Toronto, Ontario, M5S 1A8, Canada
pai@hera.med.utoronto.ca

Abstract This review aims to use the results of an approach combining crystallographic structure analysis with mutational studies as a framework for the various mechanistic proposals advanced in attempts to explain the astonishing acceleration rates displayed by orotidine 5′-monophosphate decarboxylase, the most proficient enzyme known. Special emphasis is placed on the contributions of active site amino acids to the selection and binding of substrate, product, and inhibitors as well as on the identification of alternative binding modes. Finally, a dynamic mechanism is proposed in which an increase in the strength of substrate binding and its catalytic conversion to product progress in parallel.

Keywords X-ray crystallography · Crystal structures · Alternative binding modes · Inhibitors · Electrostatic strain · Active site mutants · Dynamic reaction mechanism · Escherichia coli · Bacillus subtilis · Methanobacterium thermoautotrophicum · Yeast

Abbreviations

ODCase	Orotidine 5′-monophosphate decarboxylase
TIM	Triosephosphate isomerase
r.m.s.	Root-mean-square
BMP	6-hydroxyuridine 5′-mono-phosphate
CMP	Cytidine 5′-monophosphate
OMP	Orotidine 5′-monophosphate
UMP	Uridine 5′-monophosphate
XMP	Xanthine 5′-monophosphate

1
Overview of Structural Analyses of ODCases from Various Sources

Based on measurements of kinetic parameters and isotope effects, model reaction studies, and theoretical calculations, a surprisingly large number of hypotheses have been put forward to account for the enormous rate accelerations achieved by orotidine monophosphate decarboxylases (ODCases) [1, 2, 3, 4, 5, 6]. Until quite recently, however, there was no structural information available that could serve as a framework for mechanistic discussions. For distinguishing different proposals and for integrating the results of biochemical and mutational studies, such knowledge is a crucial prerequisite. For example, a lysine residue had been identified as essential for catalysis but its location in relation to the substrate was unknown [7]. One could only speculate about the exact nature of the substrate's environment [3] and there were arguments exchanged for and against the mandatory involvement of a divalent ion in catalysis [8, 9, 10].

Five years after Wolfenden and collaborators, pointing to the extraordinary proficiency of ODCase, created renewed interest in this enzyme [11], a surge of crystallographic results from four different laboratories provided the crystal structures of the ODCases produced by the thermophilic archaeon *Methanobacterium thermoautotrophicum*, baker's yeast, *Bacillus subtilis*, and *Escherichia coli*. The protein molecules were either in a ligand-free form or complexed to the tight inhibitors 6-azaUMP and 6-hydroxyuridine 5′-monophosphate (BMP) as well as to the product UMP [12, 13, 14, 15]. Not unexpectedly, these structures established ODCase as a member of the large family of TIM (triosephosphate isomerase)-barrel fold proteins. However, there were several surprises, too. The O2 atom of the substrate had been widely predicted as a good candidate to interact with the catalytic lysine (Lys72 in *Methanobacterium thermoautotrophicum*[1]). In contrast to those predictions, the latter was found close to C6, on the side opposite to

[1] If not stated otherwise, the residue numbers of the enzyme from *Methanobacterium thermoautotrophicum* are used in this review.

O2. The lysine side chain is held in this position by two flanking aspartate residues. Together with a second lysine (Lys42), these four residues form a charge-network that is highly conserved in all ODCases, both in primary sequence and in its three-dimensional orientation.

The crystallographic results helped to rule out some of the mechanistic proposals. For example, no metal ions were found at or near the active site in any of the ODCase structures making it very improbable that such ions could play an essential mechanistic role. The residues closest to C5 are the hydrophobic amino acids Met126, Leu123 and Val155. With no nucleophilic residue nearby, covalent catalysis, supposed to involve a bond between a protein residue and position C5 of the substrate base, can also be ruled out. Disappointingly, however, the structures did not offer any unequivocal choice of catalytic mechanism. Although the ODCases from the four sources mentioned above all displayed essentially identical active site architectures, each of the four reporting laboratories presented their own and distinct version of a reaction mechanism [12, 13, 14, 15].

Further efforts involving structural and kinetic experiments and making use of various mutants of ODCase have led to an improved understanding of the active site, even though it is still not possible to describe the exact chemical steps of the catalytic mechanism with complete certainty. In this review, the presently known crystal structures of ODCase, native as well as mutant forms, in their ligand-free form and in complex with various inhibitors will be discussed. The description will focus on the *M. thermoautotrophicum* ODCase because most of the structural work involves this enzyme; also, it is the one the authors are most familiar with (and prejudiced towards).

2 Structural Studies of Native *M. thermoautotrophicum* ODCase

2.1 Description of the Overall Structure

The ODCase monomer adopts a triosephosphate isomerase (TIM) fold, with α-helices surrounding an eight-stranded β-barrel (Fig. 1).

All highly conserved residues are located in the larger opening of the barrel. Both the extreme N- and C-termini remain disordered above the smaller opening on the opposite side of the barrel. The functional enzyme is a homodimer with two identical active sites. At the dimer interface, which takes up approximately 18% of each monomer's surface area, both ionic and hydrophobic interactions can be found. The interface traps a peculiar string of water molecules, which connect the two active sites (Fig. 2).

The oxygen atoms of these water molecules display some of the lowest temperature factors of all atoms in the *M. thermoautotrophicum* ODCase

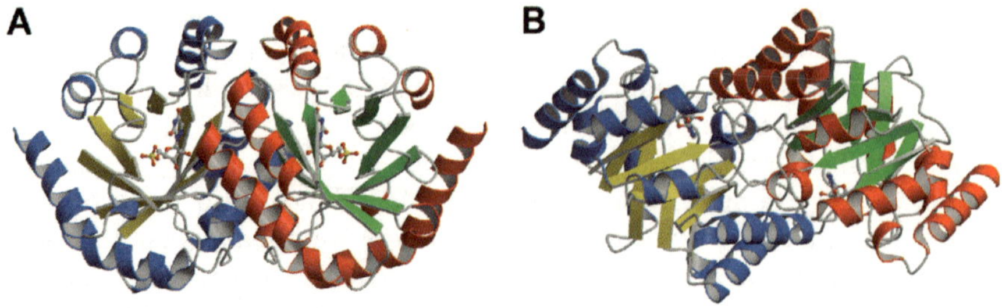

Fig. 1a,b Cartoon representation of the dimer of *M. thermoautotrophicum* ODCase. One monomer is in *red and green*, the other one in *blue and yellow*. Helices are displayed as ribbons and strands of β-sheet as arrows. The 6-azaUMP inhibitor molecules (one per monomer) are shown in ball-and-stick representation. **a** View perpendicular to the two-fold rotation axis relating the two monomers. **b** Side view, rotated by 90° compared to **a**, now looking down the rotation axis

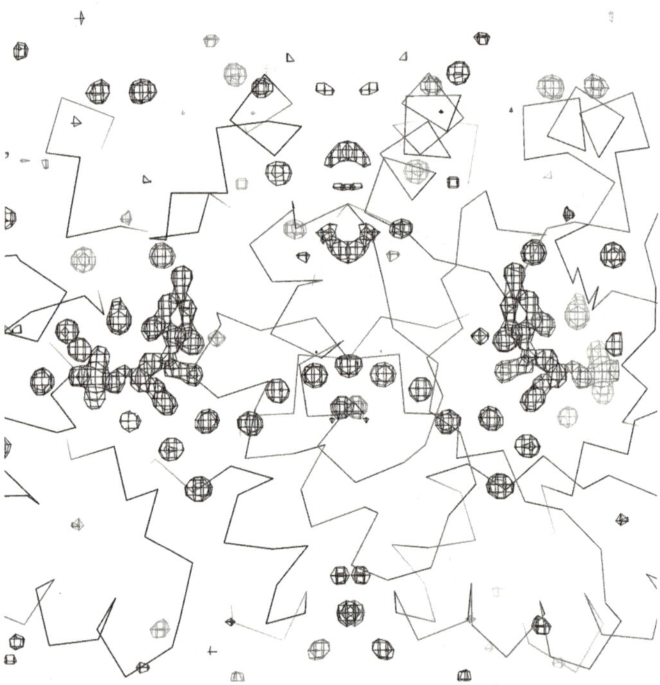

Fig. 2 F_o–F_c difference electron density map showing the two BMP ligands (one in each of the momomers) and a chain of water molecules connecting the two active sites across the dimer interface

```
                    ....|....|....|....|....|....|....|....|....|....|....|....|....|....|
B.subtilis      1   ..........................MKNNLPIIALDFASAEETLAFLAPFQQEPLFVKVGMELFYQEGP 44
E.coli          1   ...............MTLTASSSSRAVTNSPVVVALDYHNRDDALAFVDKIDPRDCRLKVGKEMFTLFGP 55
S.cerevisiae    1   MSKATYKERAATHPSPVAAKLFNIMHEKQTNLCASLDVRTTKELLELVEALGPKICLLKTHVDILTDFSM 70
M.thermoauto.   1   .................MRSRRVDVMDVMNRLILAMDLMNRDDALRVTGEVREYIDTVKIGYPLVLSEGM 53

                    ....|....|....|....|....|....|....|....|....|....|....|....|....|....|
B.subtilis     44   ..SIVKQLKERN...CELFLDLKLHDIPTTVNKAMKRLASL...GVDLVNVHAAGGKKMMQAALEGLEEG 106
E.coli         55   ..QFVRELQQRG...FDIFLDLKFHDIPNTAAHAVAAAADL...GVWMVNVHASGGARMMTAAREALVPF 117
S.cerevisiae   71   EGTVKPLKALSAKYNFLLFEDRKFADIGNTVKLQYSAGVYRIAEWADITNAHGVVGPGIVSGLKQAAEEV 140
M.thermoauto.  53   ..DIIAEFRKRFG..CRIIADFKVADIPETNEKICRATFKA...GADAIIVHGFRGADSVRACLNVAEEM 116

                    ....|....|....|....|....|....|....|....|....|....|....|....|....|....|
B.subtilis    107   TPAGKKRPSLIAVTQLTSTSEQIMKDELLIEKSLIDTVVHYSKQAEESGLD..GVVCSVHEAKAIYQAVS 174
E.coli        118   GKDA...PLLIAVTVLTSMEASDLVDLGMTLSPAD.YAERLAALTQKCGLD..GVVCSAQEAVRFKQVFG 181
S.cerevisiae  141   TKEP...RGLLMLAELSCKGSLATGE..........YTKGTVDIAKSDKDFVIGFIAQRDMGGRDEGYD. 196
M.thermoauto. 117   G......REVFLLTEMSHPGAEMFIQG.........AADEIARMGVDLGVK..NYVGPSTRPERLSRLRE 169

                    ....|....|....|....|....|....|....|....|....|....|....|....|....|....|
B.subtilis    175   PS...FLTVTPGIRMSEDA.ANDQVRVATPAIAREKGSSAIVVGRSITKAE.DPVKAYKAVRLEWEGIKS 239
E.coli        182   QE...FKLVTPGIRPQGS.EAGDQRRIMTPEQALSAGVDYMVIGRPVTQSV.DPAQTLKAINASLQRSA. 245
S.cerevisiae  196   .....WLIMTPGVGLDDKGDALGQQYR.TVDDVVSTGSDIIIVGRGLFAKGRDAKVEGERYRKAGWEAYL 260
M.thermoauto. 170   IIGQDSFLISPGVGA........QGG..DPGETLRFA.DAIIVGRSIYLAD.NPAAAAAGIIESIKDLLN 227

                    ....|.
B.subtilis    239   ......  239
E.coli        245   ......  245
S.cerevisiae  261   RRCGQQN 267
M.thermoauto. 228   P.....  228
```

Fig. 3 Structure-guided multiple sequence alignment of the four ODCases of known molecular structure. The fully conserved catalytic residues are highlighted in *black*. Other residues involved in inhibitor binding (Asp20, Ser127, Gln185, Thr79^{B}) are highlighted in *grey*. The residues lining the postulated temporary binding pocket for CO_2 are highlighted in *light grey*. The phosphate-binding loop is *boxed*. *Black arrows* indicate *β*-strands and *grey bars α* helices

molecule, indicative of very tight and rigid binding. Only some of these structural water molecules can be found in the crystal structures of ODCases from other sources, possibly due to the lower resolutions of the respective analyses. In some of the crystal forms investigated, the two-fold rotation axis relating one monomer to the other aligns with a crystallographic two-fold axis requiring identity in structure between the two subunits. In other crystal forms, in which the complete dimer is part of the asymmetric unit of the crystal and therefore no systematic restrictions on structural similarity exist, the structural differences between the two monomers nevertheless remain small, indicating an intrinsically rigid conformation of the subunits. For example, in the 6-azaUMP complex of native ODCase, the r.m.s.d between 211 corresponding C_α-atoms (residues 14 to 224) is only 0.3 Å.

Although the overall structures of the four different ODCases are very similar, the yeast protein seems to incorporate somewhat larger changes when compared to the three prokaryotic enzymes, including two additional helices at its N-terminus. Most of the variations are found in the loop regions and the outer helices. The thermophilic nature of the *M. thermoautotrophicum* enzyme might be responsible for its relatively short connections between elements of secondary structure, which create a more compact appearance; the corresponding loops for mesophiles are much longer (Fig. 3).

In the ligand-free structures, several of these loops are disordered; they acquire a rigid conformation only upon binding of a ligand [13, 16].

2.2 Interactions between Various Inhibitors and the Enzyme's Active Site

The two strong inhibitors, 6-azaUMP ($K_i=6.4\times10^{-8}$ M) [17] and BMP ($K_i=8.8\times10^{-12}$ M) [18], as well as the product UMP ($K_i=2.0\times10^{-4}$ M) [17] bind to the active site in an analogous fashion [19]. In all cases, the phosphoribose group is tightly held at the active site with the conserved Arg203 clamping down on the phosphate group (Fig. 4).

The peptide backbone amides and carbonyls of the stretch of residues on both sides of Arg203 and of those in the loop formed by amino acids 181–187 (GVGAQGG) also interact with the phosphate group, most of them indirectly through water molecules. These water molecules are found reproducibly at the same positions in most of the structures determined for the various *M. thermoautotrophicum* enzyme complexes despite drastic changes in the respective crystallographic environments.

For each of the ligands 6-azaUMP, BMP, and UMP, the ribose ring is in the 2′ *endo* conformation and the base in the *syn* orientation. Even though

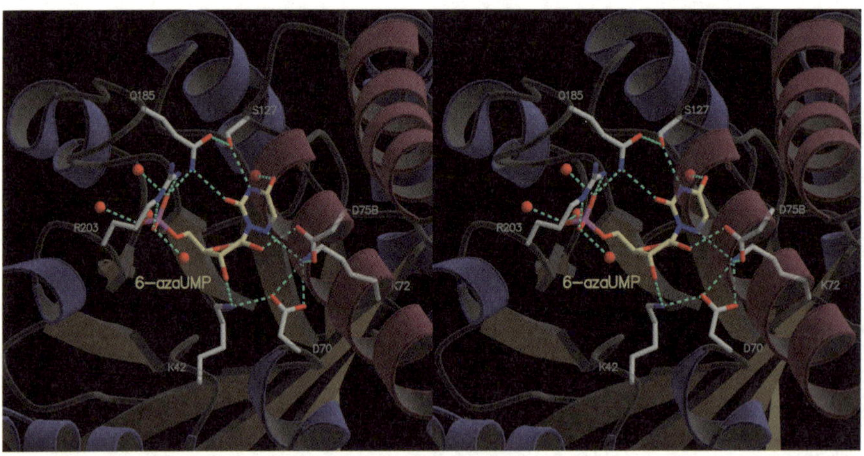

Fig. 4 Stereo representation of the interactions between active site residues in native *M. thermoautotrophicum* ODCase and the inhibitor 6-azaUMP. Several tightly bound water molecules are also shown. *Blue* helices and *yellow* β-sheets belong to one monomer, *red* helices belong to the second monomer

the energetically most favorable conformations of 6-azaUMP and UMP in solution have been identified as 3′ *endo* and *anti* [20], when bound to ODCase, the C2 oxygen of their base prefers to share a proton donor with the phosphate group. In analogy, we propose that BMP is in the same conformation although, even at a resolution of 1.5 Å, it is difficult to distinguish the two orientations of the base, which put either the C2-or the C6-oxygen close to Gln185, the regular binding partner of the C2-oxygen. Each of the ribose hydroxyls is held in place by two hydrogen bonds; Asp20 and Lys42 lock onto the 3′-OH group whilst Asp75^B and Thr79^B interact with 2′-OH. Asp75^B and Thr79^B are residues contributed by the other monomer. They are located on a loop and the following β-sheet strand, respectively. This interaction between the subunits seems to bring the two monomers closer together, with the outer helices moving slightly towards the dimer interface. This movement is much less pronounced in the *M. thermoautotrophicum* enzyme than in the mesophilic enzymes [16]. The r.m.s. deviation between the C_α atoms of the complexed and ligand-free dimers is less than 0.9 Å. Ile76 is another amino acid that is contributed to the active site by the second monomer; it is highly conserved and forms part of the hydrophobic environment.

When compared to the tight binding of the phosphate and ribose groups, the base of the ligand interacts rather loosely with the enzyme active site. The only position on the base that is specifically recognized is N3, which forms a direct hydrogen bond to the Ser127 side chain hydroxyl. A water molecule bridges the C4 oxygen to the backbone carbonyl of Glu125 and to the backbone amide of Ser127. As mentioned before, the C2 oxygen shares a hydrogen bond donor with the phosphate group, which can be a water molecule or the side chain amide of Gln185.

The pyrimidine ring is surrounded by a number of hydrophobic residues arranged at the barrel opening. The chemical identities of these residues are not strictly conserved, but their hydrophobic character is (Fig. 3). The four residues Lys42, Asp70, Lys72, and Asp75^B, which are completely conserved in all known ODCase sequences, are arranged in a charge network flanking one side of the pyrimidine ring as well as the ribose hydroxyls. The two outermost residues of the network, Lys42 and Asp75^B, associate with the 3′- and 2′-hydroxyls of the ribose ring, respectively. The two center residues, the catalytic Lys72 and Asp70, are placed close to the actual reaction site, C6. This unusual network of electrostatic interactions set against a hydrophobic background is poised to play an important role in catalysis (Fig. 4).

When the various inhibitor complexes are compared to each other, little change in the protein's structure is observed; significant structural rearrangements are restricted to the relative orientations of the dimers (r.m.s.d. of 0.5 Å for 418 C_α of the dimer overall and of 0.2 Å for 50 backbone and C_β atoms of the conserved active site residues Asp20, Lys42, Asp70, Lys72, Ser127, Gln185, Arg203, Asp75^B, Ile76^B and Thr79^B). Binding of UMP, 6-aza-

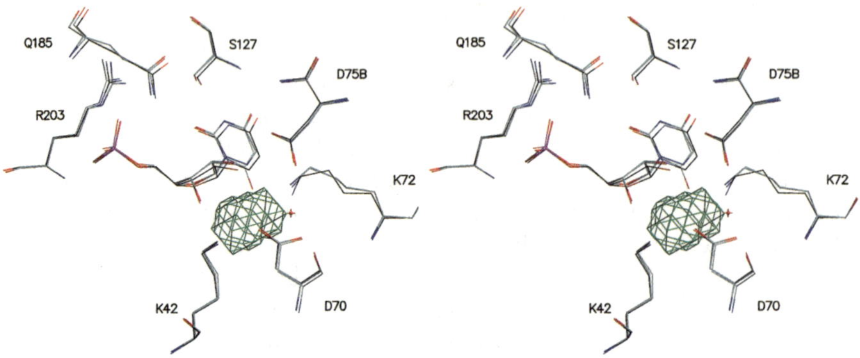

Fig. 5 Superposition (in stereo) of 6-azaUMP, UMP and BMP bound to the active site of native *M. thermoautotrophicum* ODCase. Carbon atoms of the BMP and 6-azaUMP complexes are colored in *grey* and those of the UMP complex are in *black*. The *green* contour (accessible area calculated with 1.2 Å probe radius using VOIDOO [29]) represents the cavity located next to the pyrimidine bases and proposed as a temporary binding site for the product CO_2

UMP, and even BMP, which brings with it a water molecule solvating the C6 oxygen and forming a hydrogen bond to Asp70, leaves a void in the active site next to the pyrimidine ring. This cavity is created by the displacement of water molecules, which bind to the charged residues in the active site of the freeform (Fig. 5).

The cavity's location and shape as well as its partly hydrophilic, partly hydrophobic character make it well suited to act as the first intermediary binding site of the product CO_2, immediately after bond breakage. The binding site for the base of the ligand is spatially quite restricted; it seems to be severely limited in its ability to accept substituted rings requiring much more space than the bases of UMP or 6-azaUMP. For the substrate OMP, this situation would even be aggravated because its binding to an unmodified active site would place two negative charges in very close proximity [12]. This has led us to speculate that once OMP reaches the same binding position as 6-azaUMP in the active site, the C6-carboxylate bond has already been broken or it is at least severely distorted with the carboxylate group tilted strongly towards the location of the cavity.

BMP (K_i=8.8×10^{-12} M) [18] and 6-azaUMP (K_i=6.4×10^{-8} M) [17] are very tight binding inhibitors whereas the product UMP is a rather weak one (K_i=2.0×10^{-4} M) [17]. It is interesting to note that there are roughly four orders of magnitude difference between each inhibition constant, which seems to correlate with the amount of negative charge close to position 6 of the pyrimidine ring, the location where such a charge will develop in the carbanion intermediate: the C6 oxygen of BMP is fully ionic in character, N6 of 6-azaUMP carries a partial charge, and C6 of UMP is neutral. The distances between these atoms and the amino head group of Lys72 reflect this, too: 2.8 Å for BMP, 3.0 Å for 6-azaUMP, and 3.9 Å for UMP. In contrast to the first two cases, C6 of UMP cannot act as a partner in ionic or hydrogen bond interactions and the lysine side chain actually curls, moving the amino head group further away (Fig. 5). In addition, nucleotides, upon binding, expel several water molecules from the crevice that accepts the ligand's base. In the UMP complex, this leaves a void whereas in the BMP complex, a water molecule is trapped contributing to the maximization of van der Waals forces.

3 Structural Studies of Mutants of *M. thermoautotrophicum* ODCase

In further efforts to probe the enzyme active site, we systematically mutated the base-recognition residues Ser127 and Gln185, the charge network residues (Asp75^{B}, Lys42, Asp70, and Lys72) as well as the phosphate-binding residue Arg203 and the loop ^{184}AQGG^{187} [19, 21]. There is no major movement of the backbone in any of the nine mutants whose structures have been analyzed. In some of them, notably D75^{B}N, one observes slight local shifts of side chains close to the mutation. In the active site, all residues appear to sit in well-defined minimum energy positions, even in the presence of mutational disturbances. The only exception is the D70G mutant, in which Lys72 becomes disordered in the absence of its neighboring side chain. The r.m.s. deviation for 45 main chain and C_β atoms of conserved residues in the active sites of our nine mutants is less than 0.2 Å. The protein molecules are complexed with three different inhibitors, product, and substrate, and are all crystallized from solutions of different pH and precipitant conditions. This rigidity of the active site had already been recognized by the close superposition of corresponding atoms in ODCases from four sources, each one crystallizing in a different space group.

3.1 Base-Recognition Mutants

In crystallographic analyses of complexes of native ODCase, we found only two protein residues that directly interacted with the pyrimidine base: Ser127 and Gln185. The side chain of Ser127 forms a hydrogen bond to N3, its main chain amide links to O4. While this residue is not strictly conserved in ODCases (Thr in *B. subtilis* and *E. coli*), its side chain hydroxyl function is. Its role in base-recognition seems to be more important than that of the highly conserved Gln185, which can be replaced by a water molecule. In different space groups, Gln185 shows multiple conformations, reflecting the flexibility of the phosphate-binding loop. Indeed, the S127A mutation result-

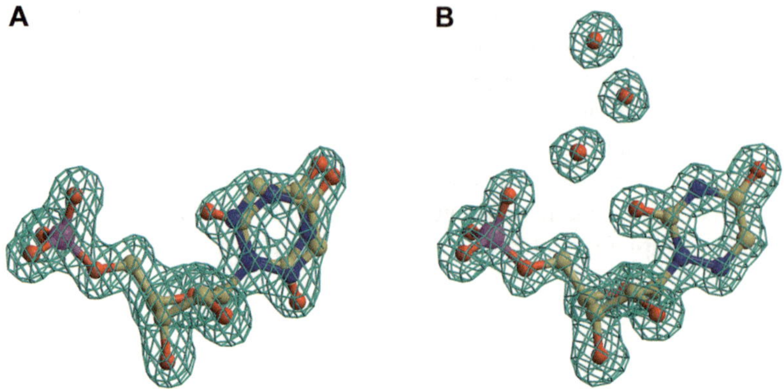

Fig. 6a,b Electron density representing the inhibitor of 6-azaUMP in the active sites of base-recognition mutants. **a** shows the dual conformations adopted by the pyrimidine ring in S127A ODCase. **b** In the Q185A mutant, a chain of water molecules replaces the glutamine side chain

ed in changes in the conformation of the bound inhibitor whereas nothing like it was found for the Q185A mutant. These structural results nicely mirror the excellent kinetic measurements performed on the yeast enzyme [22].

To be more specific, the Ser127 hydroxyl to N3 interaction locks the 6-azaUMP nucleotide in its energetically less favorable *syn* conformation. If a mutation removes this contact, both *syn* and *anti* conformations are equally populated as seen in the 6-azaUMP complex of S127A ODCase (Fig. 6a).

Since OMP carries a much larger substituent on C6, it is not likely that the nucleotide could adopt a conformation that would place the carboxylate on top of the sugar ring. Nevertheless, the N3-Ser127 interaction contributes to substrate specificity and to the locking in of the proper orientation.

On the other hand, when Gln185 is replaced by alanine, the void created by the missing side chain atoms is filled with a row of well-ordered water molecules (Fig. 6b).

One of them occupies the position of the side chain amide nitrogen, donating hydrogen bonds to both the 5′-phosphate and the O2 atom of the base. This glutamine residue sits in the middle of the sequence that forms a mobile loop, which is only structured when clamping down on a 5′-phosphate group.

3.2
Charge Network Mutants

Stretched out along one side of the substrate, amino acids Lys42, Asp70, Lys72, and Asp75^{B} closely interact with each other, forming an intricate charge network. Strict conservation [23] as well as conspicuous location point to their crucial, although still quite controversial, role for the enzyme's function [12, 13, 14, 15, 27, 28]). As these residues all interact very closely with each other, pinpointing their individual roles is quite difficult. Nevertheless, single-site mutants were constructed in an attempt to dissect the contributions of the various side chains to the catalytic mechanism.

Whether it is the central position proposed for Asp70 in the electrostatic stress mechanism [12] or the role of Lys72 as proton-donor to the carbanion intermediate [24], it is obvious that these two residues are the most important ones in catalysis. Nevertheless, all Asp70 as well as Lys72 single-site mutants of *M. thermoautotrophicum* ODCase that were generated still retained some enzymatic activity (see below), a clear indication that it is not a sole residue carrying all of the catalytic load [21].

For Asp70 to exert sufficient levels of "stress" on the substrate carboxylate group to cause its release as carbon dioxide, the side chain should be negatively charged. This would then dictate that the other three members of the network also carry charges. In an alternative mechanistic proposal, OMP picks up a proton as it is forced into the active site and shares this proton with the side chain of Asp70 in a low-energy hydrogen bond [15]. The very close approach between the amino acids of the charge network and their balancing effects on each other then imply that all four of them are uncharged. Direct experimental evidence about the exact positioning of protons will have to wait for the results of ultra-high resolution X-ray or neutron diffraction studies. In addition to the potential charge–charge interaction, one should expect a steric conflict between the OMP carboxylate and the Asp70 side chain.

In an attempt to trap an intact substrate molecule in the active site, the mutant D70A was co-crystallized with OMP. However, the product UMP was found in the crystal structure along with a Cl^- ion, which occupied space vacated by the missing aspartate side chain and was held in place by its neighbors Lys72 and Lys42 (Fig. 7b).

It is reasonable to assume that under the conditions of the crystallization experiment, i.e., very high protein concentration and long durations, even a low residual enzymatic activity could still be sufficient to cause the conversion of OMP to UMP before crystal nucleation occurs. One could argue that the Cl^- ions, present in the crystallization buffer and picked up by the mutant enzyme, could at least partially substitute for a negatively charged side chain. Given this interesting, but still somewhat disappointing, result together with kinetic tests that indicated weak but detectable enzymatic activity also for the D70G and D70 N mutants of *M. thermoautotrophicum* ODCase (unpublished results), 6-azaUMP was chosen as the complexing ligand for their structural analyses. In the D70G mutant, there is enough space created in the active site to allow water molecules to successfully compete with O2, leading to an equal distribution between the two stable conformers of the base (Fig. 7a).

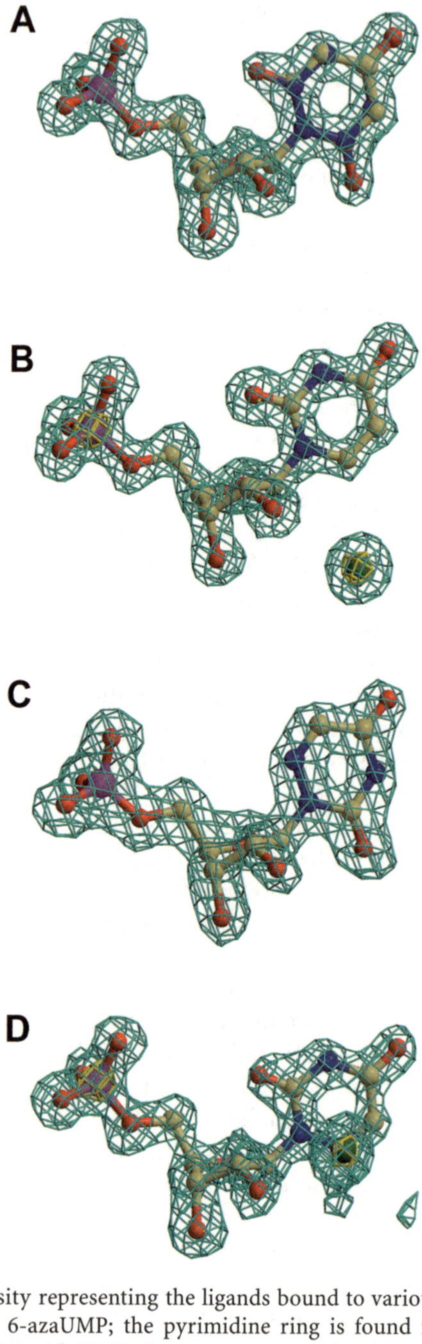

Fig. 7a–d Electron density representing the ligands bound to various active site aspartate mutants. **a** D70G with 6-azaUMP; the pyrimidine ring is found in two orientations of equal occupancy. **b** D70A with UMP; the mutant enzyme has incorporated a chloride ion from the crystallization solution. **c** D70N with 6azaUMP; in this complex, all 6azauracil rings are rotated by 180°. **d** D75[B]N with 6azaUMP; again, a chloride ion substitutes for the negative charge of an aspartate side chain lost by the mutation

These water molecules cannot efficiently fix the side chain of Lys72, which therefore becomes quite mobile. The presence of a partial negative charge on N6 seems to prevent incorporation of a large Cl^- ion into the active site of D70 N ODCase. Now, all molecules have their pyrimidine rings rotated by 180° placing the more electron-rich O2 atom closer to the lysine residues (Fig. 7c).

When the two flanking residues of the charge network, $Asp75^B$ and Lys42, are investigated, one finds that the $D75^BN$ mutant, too, recruits a Cl^- ion to the active site to preserve the negative charge (Fig. 7d).

The chloride is coordinated to the ribose 2′-OH, Lys72, and the backbone amide of Ile76, the same partners engaged by the aspartate in nucleotide complexes of native protein. As a D to N mutation does not create a void, the presence of the chloride ion is accommodated by the asparagine side chain swinging out of the active site leading to the largest movement observed in all of the mutants tested.

Compared to those of the aspartate mutants, the results of changing the lysine residues Lys42 and Lys72 appear to be less varied. In both cases, the void left by the removal of the lysine side chains is simply occupied by water molecules. Although kinetic experiments with yeast ODCase did not find any enzymatic activity for the mutant corresponding to K72A [22], setting the latter up for crystallization in a solution containing OMP, resulted in electron density corresponding to UMP in the base-binding site [21]. This parallels our findings with the D70A mutant and confirms that neither the negative charge on Asp75 nor the proton donating capability of Lys72 on their own is the sole driving force for the reaction.

Only when both Asp70 and Lys75 are changed into alanines, is the catalytic power of ODCase reduced sufficiently to enable the stable binding of the substrate OMP to the active site. In addition, the removal of the two side chains now allows the active site to easily accommodate the carboxylate group, which is not in plane with the pyrimidine ring (as found in all the orotidine-containing structures in the Cambridge Data Base). Interestingly, the carboxylate is slightly tilted away from the space created by the removal of side chains and points towards the cavity that has been proposed as a potential transitory binding site for the product CO_2 (Fig. 8).

The importance of the presence of a negative charge at the aspartate positions in the charge network, in a role other than as a catalyst, is exemplified in the loss of stability shown by the respective mutants. They either form inclusion bodies or are prone to precipitate; they also unfold at much lower concentrations of denaturants [21]. It would be most intriguing to know how structural integrity and catalytic efficiency co-evolved around the negative charge at Asp70.

The most probable candidate for the proton donor, required to neutralize the carbanion intermediate, is the side chain of Lys72; it is held in place by the two flanking aspartates, Asp70 and $Asp75^B$. In addition, the close inter-

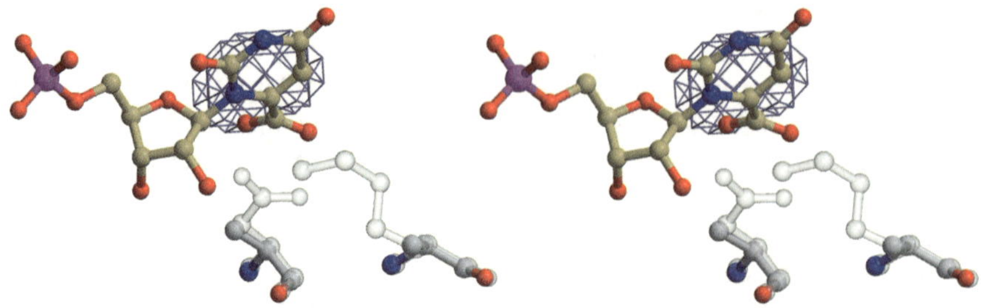

Fig. 8 Stereo representation of the substrate OMP bound to the double mutant D70A/K72A. Note how close the C6-carboxylate is located to the internal cavity (*purple density*) postulated as a temporary binding site for CO_2. The positions of the side chains of Asp70 and Lys72 in the native enzyme are shown in *transparent grey*

actions taking place between the four charged side chains should make for a rather rigid wall on their side of the base-binding crevice. Asp70, especially, is taking up a position in which it will strongly interfere, both sterically and electrostatically, with an approaching C6 carboxylate, preventing the C6 position to come into contact with the ammonium group of Lys72 before the C6–C6 carboxylate bond is broken. Set against a generally rather hydrophobic environment, movement of each one of the four residues in the charge network would probably cause accommodating changes of the others. The tight and multiple interactions between them will act to prevent such position changes. The nucleotide substrate/ligand will contribute to a stiffening of the binding site because the two end residues in the network, Lys42 and Asp75^B, also bind to the hydroxyl groups of the ribose, thereby holding the whole ligand in place. These effects make a detailed dissection of the catalytic mechanism rather difficult, with each residue performing more than one specific role. It is even harder to describe the contributions of the hydrophobic residues that help to create a unique microenvironment suitable for achieving such an enormous rate acceleration. Their importance, however, should not be disregarded and certainly not all the catalytic power of this remarkable enzyme can be attributed to either Asp70 or Lys72 alone.

3.3 ΔR203A, a Phosphate-Binding-Loop Mutant

As shown by kinetic studies for ODCase, the binding energy contributed by its interaction with the 5′-phosphate group is very important for catalysis [17]. Mutants with impaired phosphate-binding, e.g., Y217A, R235A, and the double mutant Y217A/R235A in the yeast enzyme, show dramatically reduced catalytic rates. As the *M. thermoautotrophicum* enzyme with its shorter loops is missing the residue corresponding to Tyr217, we created a mutant with reduced phosphate-binding ability by replacing the other phos-

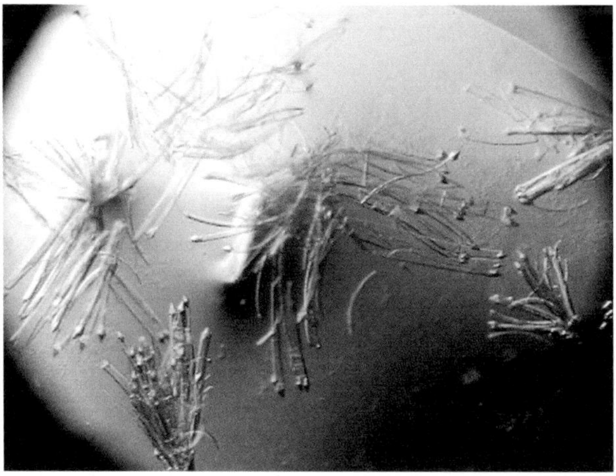

Fig. 9 Crystals of the 6-azaUMP complex of the ΔR203A mutant of *M. thermoautotrophicum* ODCase. Older crystals have bent and crystals of apo-enzyme (small diamonds) grow on their tips

phate-interacting amino acid, Arg203, with an alanine in addition to deleting the sequence Ala184–Gln185–Gly186–Gly187 in the mobile loop, amino acids that chelate the 5′-phosphate group via their main chain peptide bonds. Gly183 and Asp188 are positioned close enough in space to be connected directly without inducing undue strain on the strand of β-sheet and the α-helix, the two parts of secondary structure connected by this loop.

With a very high concentration of 6-azaUMP in the crystallization setups, complex crystals can be induced and further stabilized by addition of Mg^{2+} ions (although the metal ion cannot be identified in the resulting electron density map). After a few days, however, ligand-free crystals will nucleate on top of the deforming complex crystals and grow at their expense (Fig. 9).

Obviously, the inhibitor binds only weakly to the mutant enzyme and will diffuse out of the crystal lattice over time. This behavior can be explained when the interactions between this ligand and the mutant enzyme are analyzed in detail. Although the inhibitor occupies the exact same position as its counterpart in the native ODCase, there are only very few contacts left between nucleotide and protein matrix to hold the phosphate in place (Fig. 10).

The ribose ring makes its usual contacts with the enzyme, with two hydrogen bond links for each hydroxyl. These interactions are probably the reason why the second monomer is still drawn in closer to the active site and closes it off. The overall instability and flexibility of this complex is reflected in high B factors (35–40 Å^2) displayed by all components of the complex.

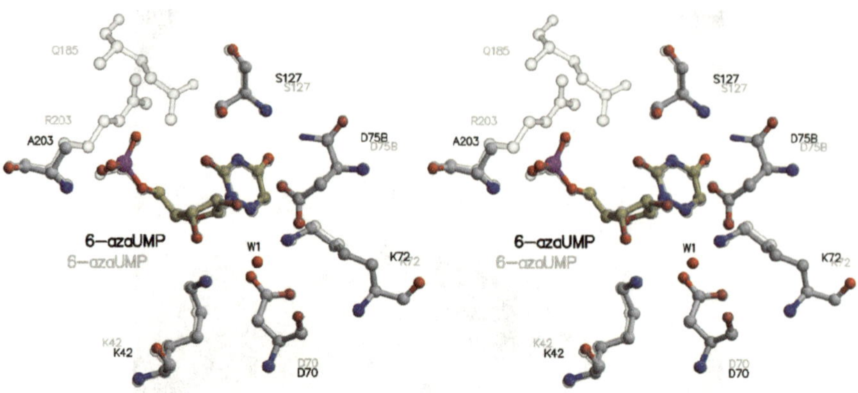

Fig. 10 Superposition (in stereo) of the active sites of the 6-azaUMP complexes of the R203A mutant (*solid colors*) and the native enzyme (*transparent grey*). Note the exact fit of the two nucleotides despite the absence of strong binding interactions in the mutant protein

4 Identification of an Alternate Binding Mode

Based on the results of the structural studies described above, we believe that the binding site of the nucleotide base is fairly small and generally quite rigid. There is only a small void left, surrounded by hydrophobic residues and located next to the six-membered ring of the base; a site which could well function as an intermediary location for the CO_2 immediately after being produced in the catalytic reaction. Given this restricted environment, it is surprising to find that the purine nucleotide XMP displays a lower inhibition constant than the product UMP.

Unfortunately, no kinetic inhibition parameters are available for CMP but it has been established that its derivative, CMP-6-carboxylate, is not accepted as a substrate by ODCase [10]. Related crystal structures alone, cannot provide an obvious explanation why this would be the case. The base of UMP, when bound to ODCase, undergoes four interactions with the protein matrix: Ser127 (probing N3), Gln185 (probing O2) as well as the backbone amide of Ser127 and a water molecule (both probing O4). Three of these internal sensors of the chemical nature of the bound nucleotide base could act on CMP as well; the backbone NH to O4 hydrogen bond found in the UMP complex is the only one that could not be formed in the CMP complex.

In an effort to shine more light onto this puzzling question, we co-crystallized native ODCase with XMP and CMP, respectively. The structural analyses of the resulting complex crystals revealed a second mode of binding. Whereas the phosphates occupy the same place as in all other complexes, the ribose groups are moved somewhat from the positions they adopt in the UMP, 6-aza-

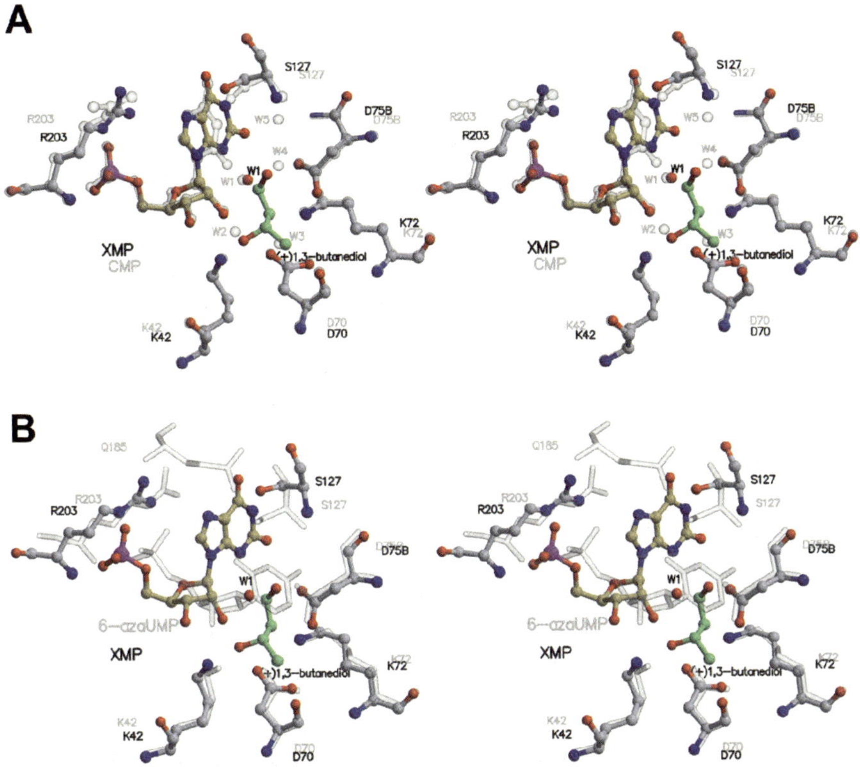

Fig. 11a,b Stereo representation of a superposition of the active sites of native *M. thermoautotrophicum* ODCase complexed with **a** XMP (*solid colors*) and CMP (*transparent grey*) and **b** XMP (*solid colors*) and 6-azaUMP (*transparent grey*), respectively

UMP, and BMP complexes. The remarkable difference, however, is the binding sites of both the purine and the pyrimidine bases. They are located outside the base-binding crevice, pointing towards the surface of the enzyme. They assume overlapping positions and display the same lower energy 3′ *endo*-pucker and *anti* conformation favored in solution (Fig. 11a) and might indicate a substrate conformation found along the reaction coordinate.

It should also be noted that in both cases the phosphate-binding loop is disordered, reminiscent of the ligand-free enzyme structure. Arg203 still holds on to the phosphate group, but less tightly, a fact which is particularly obvious in the CMP complex where its side chain is significantly more disordered in one monomer than in the other [19]. Shifted from their original positions, the ribose rings can only contact Lys42 via their hydroxyl groups. They are, however, not close enough to Asp75^B and Thr79^B to establish the two tight hydrogen bonds that are essential for changing monomer–monomer orientation.

Despite the rather dramatic shift in the position of the nucleotide bases, Ser127 is still interacting with them. It forms two hydrogen bonds to N1 and O2 of the xanthine ring. The space between the xanthine base and the residues of the charge network is bridged by a water molecule and a molecule of (+)1,3-butanediol, an additive used in the crystallization of this complex. The water molecule assumes the regular binding spot of the ribose 2′-OH group, including the formation of hydrogen bonds to the side chains of $Thr79^B$ and $Asp75^B$ (Fig. 11b).

In the absence of ligands, a first-hydration-shell water interacting with $Asp75^B$ takes up this position, indicating a strong preference for a hydroxyl function at this location. The hydrophobic chain of the butyl alcohol backs against the hydrophobic pocket while the C1 and C3 hydroxyls bind to N3 of xanthine and 2′-OH of the ribose, respectively, replacing several hydration shell waters found in the ligand-free enzyme structure. Kinetic studies did not detect any significant influence on XMP inhibition by the butyl alcohol. Its less well-defined electron density together with much increased B factors (average of 44 $Å^2$ compared to values of 21 $Å^2$ for XMP and 16 $Å^2$ for active site residues) are consistent with this finding [5].

Addition of another small organic compound, 1,2,3-heptanetriol, proved essential to obtain diffraction-quality crystals of the CMP complex. In this case, however, no electron density representing the alcohol is found. Instead, five water molecules fill the same space, one of them again at the position so favorable to hydroxyl group binding. Although the xanthine and cytosine rings superimpose in the XMP and CMP complexes, because of its smaller size, no contacts exist between the cytosine ring and ODCase. Two water molecules bridge the main chain amide of Ser127 and cytosine's O2. The major interactions that hold the CMP ligand in place are the hydrogen bonds between the two ribose hydroxyls and Lys42. The contacts between the phosphate group and the protein are weakened due to the disordered state of the phosphate-binding loop, Gly^{180} to Gly^{190}, and the increased mobility of Arg203.

5
Mechanistic Discussions

It is particularly puzzling that four hydrogen bonds to two ribose hydroxyls are able to conserve 6-azaUMP-binding to the active site of ODCase, even with most of the phosphate-protein contacts removed, e.g., in the ΔR203A mutant, while the pyrimidine ring of CMP, with all the potential atom-to-atom interactions available, is expelled from the base-binding site. It seems that at least one other selection parameter has to come into play. CMP, the high-affinity inhibitors, as well as the product and weak inhibitor UMP all have pyrimidine bases, therefore, physical shape cannot be the distinguishing factor. The vast majority of hydrogen bonds or electrostatic interactions can be formed with all of these molecules. What is different, however, are the dipole moments of the various pyrimidines. The substrate OMP, the carbanion intermediate, and the transition state analogues BMP and 6-azaUMP have the negative end of their dipole moments pointing in the direction of C6; UMP does not, nor does CMP [25]. Whereas the uracil base still contributes to the binding energy (UMP is a better inhibitor than ribose 5′-phosphate [17]), no such favorable interaction seems possible for cytosine. As all residues in and close to the active site contribute to the electrostatic environment, it becomes very difficult to assign to them very specific roles in binding or catalysis. The same arguments can be made for the substrate molecule.

The conformations of XMP and CMP may represent a transient binding mode of an OMP substrate molecule after it has docked onto the enzyme and before it is drawn deeper into the active site where it will form the carbanion intermediate. It seems that the interactions that are most important in attracting and immobilizing the substrate are the same ones that are involved in the stabilization of the transition state [24, 26]. One can imagine the progress of the reaction as a process in which the continuous increase in the strength of hydrogen bonds, electrostatic interactions, and hydrophobic forces, accompanied by a change of monomer–monomer orientation, leads to a decrease in activation energy sufficient to allow bond breakage to occur.

6
Future Directions

Despite the combined efforts of several groups of biochemists and structural biologists, elucidation of the exact chemical mechanism employed by ODCase to achieve its most remarkable rate acceleration has proven elusive. Several mechanistic proposals have been refuted but new ones have emerged and need to be tested [27, 28]. As a laboratory specializing in structural biology, we have decided to apply the technique of time-resolved crystallography in the hope of being able to watch the actual transformation of OMP into UMP "as it happens". If successful, this should be a major step forward towards solving the enigma of ODCase catalysis.

References

1. Ondetti MA, Cushman DW, Sabo EF, Cheung HS (1979) Design of active-site-directed reversible inhibitors of exopeptidases. Elsevier, Holland
2. Beak P, Siegel B (1976) J Am Chem Soc 98:3601
3. Lee JK, Houk KN (1997) Science 276:942

4. Silverman RB, Groziak MP (1982) J Am Chem Soc 104:6434
5. Acheson SA, Bell JB, Jones ME, Wolfenden R (1990) Biochemistry 29:3198
6. Smiley JA, Paneth P, O'Leary MH, Bell JB, Jones ME (1991) Biochemistry 30:6216
7. Smiley JA, Jones ME (1992) Biochemistry 31:12162
8. Cui W, DeWitt JG, Miller SM, Wu W (1999) Biochem Biophys Res Commun 259:133
9. Miller BG, Smiley JA, Short SA, Wolfenden R (1999) J Biol Chem 274:23841
10. Smiley JA, Saleh L (1999) Bioorg Chem 27:297
11. Radzicka A, Wolfenden R (1995) Science 267:90
12. Wu N, Mo Y, Gao J, Pai EF (2000b) Proc Natl Acad Sci USA 97:2017
13. Miller BG, Hassell AM, Wolfenden R, Milburn MV, Short SA (2000a) Proc Natl Acad Sci USA 97:2011
14. Appleby TC, Kinsland C, Begley TP, Ealick SE (2000) Proc Natl Acad Sci USA 97:2005
15. Harris P, Navarro Poulsen JC, Jensen KF, Larsen S (2000) Biochemistry 39:4217
16. Harris P, Poulsen JC, Jensen KF, Larsen S (2002) J Mol Biol 318:1019
17. Miller BG, Snider MJ, Short SA, Wolfenden R (2000b) Biochemistry 39:8113
18. Levine HL, Brody RS, Westheimer FH (1980) Biochemistry 19:4993
19. Wu N, Pai EF (2002b) J Biol Chem 277:28080
20. Saenger W, Suck D, Knappenberg M, Dirkx J (1979) Biopolymers 18:2015
21. Wu N, Gillon W, Pai EF (2002a) Biochemistry 41:4002
22. Miller BG, Snider MJ, Wolfenden R, Short SA (2001a) J Biol Chem 276:15174
23. Traut TW, Temple BR (2000) J Biol Chem 275:28675
24. Miller BG, Wolfenden R (2002) Annu Rev Biochem 71:847
25. Saenger W (1973) Angew Chem Int Edn 12:591
26. Miller BG, Butterfoss GL, Short SA, Wolfenden R (2001b) Biochemistry 40:6227
27. Warshel A, Florián J, Strajbl M, Villà J (2001) Chembiochem 2:109
28. Houk KN, Lee JK, Tantillo DJ, Bahmanyar S, Hietbrink BN (2001) Chembiochem 2:113
29. Kleywegt GJ, Jones TA (1994) Acta Crystallogr D Biol Crystallogr 50:178

Top Curr Chem (2004) 238:43–62
DOI 10.1007/b94538

Insight into the Catalytic Mechanism of Orotidine 5′-Phosphate Decarboxylase from Crystallography and Mutagenesis

Brian G. Miller (✉)

Department of Biochemistry, University of Wisconsin, Madison, WI 53706–1544, USA
bgmiller@biochem.wisc.edu

Abstract With a half-time of 78 million years at room temperature, the spontaneous decarboxylation of orotidine 5′-phosphate (OMP) is among the slowest biological reactions to be described. Orotidine 5′-phosphate decarboxylase (ODCase), the enzyme responsible for catalyzing this difficult transformation, performs its task without the assistance of metals or small molecule cofactors. The crystal structures of ODCase from yeast and three prokaryotes have recently been determined. These structures reveal a quartet of charged residues located within the active site, near where the substrate's reactive carboxyl group is expected to bind. The results of site-directed mutagenesis of individual active site residues have demonstrated the importance of this charged network to catalysis by yeast ODCase. Large connectivity effects, operational primarily in the transition state for decarboxylation, are observed when individual binding determinants of the substrate are chemically removed. In particular, the phosphoribosyl group of substrate OMP appears to be one feature by which the enzyme distinguishes the substrate in the ground state from the altered substrate in the transition state. Current proposals for the mechanism of enzymatic decarboxylation are discussed in light of recent computational, structural, and mutagenesis studies.

Keywords Orotidine 5′-phosphate decarboxylase · Catalytic proficiency · Transition state stabilization · Site-directed mutagenesis · Connectivity effects

Abbreviations and Symbols

ODCase	Orotidine 5′-phosphate decarboxylase
OPRTase	Orotate phosphoribosyltransferase
OMP	Orotidine 5′-phosphate
UMP	Uridine 5′-phosphate
BMP	6-hydroxyuridine 5′-phosphate
TIM	Triosephosphate isomerase
Å	Angstrom
k_{cat}	Turnover number
K_m	Michaelis constant
k_{cat}/K_m	Catalytic efficiency
K_{tx}	Dissociation constant for the altered substrate in the transition state
ITC	Isothermal titration calorimetry

1 Introduction

Orotidine 5′-phosphate decarboxylase (ODCase, E. C. 4.1.1.23) catalyzes the decarboxylation of orotidine 5′-phosphate (OMP) to form uridine 5′-phosphate in the sixth and final step of pyrimidine biosynthesis (Fig. 1) [1]. The discovery of ODCase in 1954 followed the identification, three years earlier, of orotic acid as the metabolic precursor of nucleic acids [2, 3]. ODCase is a distinct, monofunctional polypeptide in bacteria and fungi, whereas in mammals it combines with orotate phosphoribosyltransferase (OPRTase) to form the bifunctional enzyme UMP synthase. Human deficiencies in either OPRTase or ODCase activity result in an autosomal recessive disorder called hereditary orotic aciduria [4]. The disease is characterized by depleted levels of pyrimidine nucleotides in the blood and by the appearance of crystalline

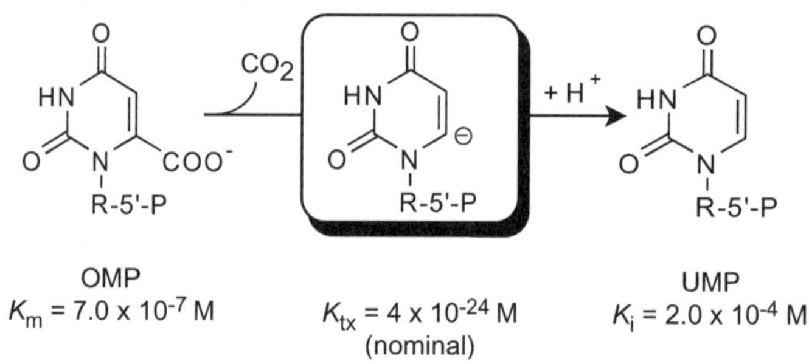

Fig. 1 The reaction catalyzed by ODCase, showing the structure of a possible carbanionic intermediate formed during decarboxylation. R-5′-P represents a 5′-phosphoribosyl group

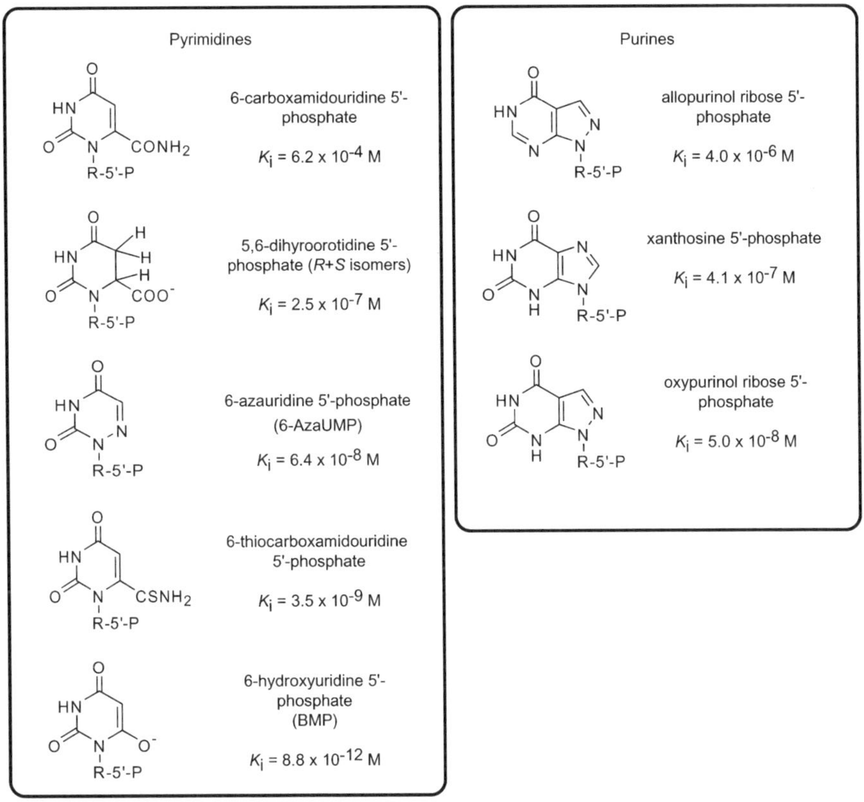

Fig. 2 The structures and binding affinities of eight competitive inhibitors of yeast ODCase. R-5′-P represents a 5′-phosphoribosyl group

orotic acid in the urine of patients. Orotic aciduria is rarely life threatening and readily treatable with uridine based dietary supplements.

Due to its role in de novo pyrimidine biosynthesis, ODCase proved a logical target for drug development efforts throughout the latter half of the 20th century. Compounds that inhibit ODCase activity were found to impair DNA synthesis in all cell types, and several were tested as potential anti-cancer agents [5, 6]. Although the use of ODCase inhibitors as chemotherapeutics proved largely unsuccessful, these efforts did result in the identification and characterization of a variety of potent inhibitors of the human enzyme (Fig. 2). Surprisingly, several effective inhibitors of ODCase activity were based on a purine rather than a pyrimidine nucleotide scaffold. Despite the attention given to the identification of ODCase inhibitors during the 1960s and 1970s, the mechanism of enzymatic decarboxylation remained a mystery.

The decarboxylation reaction catalyzed by ODCase is an extremely difficult transformation, with an uncatalyzed half-time of 78 million years in neutral, aqueous solutions at 25 °C [7]. The intrinsic sluggishness of the spontaneous reaction establishes the loss of CO_2 from OMP among the slowest biological reactions to be measured. Only the spontaneous decarboxylation of the amino acid glycine, with a half-time of 1.1 billion years, appears to occur at a slower rate under ambient temperatures [8]. Unlike amino acid decarboxylases, ODCase appears to function in the absence of metals or small molecule organic cofactors [9, 10, 11], despite the catalytic advantages offered by such prosthetic groups[1]. This fact suggests that ODCase operates by a novel chemical mechanism, and has led to a variety of experiments aimed at elucidating the catalytic forces employed by this proficient catalyst.

2 Proposed Mechanisms of Enzymatic Decarboxylation

The primary function of a decarboxylation reaction catalyst is to assist in the delocalization of negative charge that accompanies the release of CO_2 from the substrate molecule. Often this is achieved by promoting a redistribution of electrons within the substrate itself. The substrate for the ODCase reaction, however, lacks a π orbital into which electrons from the carbanionic intermediate can be delocalized. For this reason, the decarboxylation of OMP presents a special mechanistic challenge to the enzyme. To understand how ODCase might aid in delocalizing charge in the transition state for decarboxylation, Beak and Siegel conducted studies on the loss of CO_2 from 1,3-dimethylorotate at elevated temperatures under several different conditions [13, 14]. Decarboxylation of 1,3-dimethylorotate in sulfolane suggested that the nonenzymatic reaction may be initiated by the formation of the 1,3-dimethylorotate zwitterion (Fig. 3a), in a process that involves generation of a positively charged nitrogen ylide at N-1. Beak and Siegel hypothesized that this ylide could function to stabilize the development of negative charge at the adjacent C-6 atom during decarboxylation. Based on the results of their 1,3-dimethylorotate studies, the authors postulated that the ODCase active site includes an acidic residue that could protonate the O-2 atom of substrate OMP. Protonation at O-2 prior to decarboxylation is expected to shift the equilibrium in favor of the zwitterionic species, thereby promoting release of CO_2.

Several years after Beak and Siegel's proposal, Westheimer and coworkers synthesized 6-hydroxyuridine 5′-phosphate (BMP), a potent inhibitor (K_i= 8.8×10^{-12} M) of yeast ODCase [15]. The structure of this competitive inhibi-

[1] For example, pyridoxal phosphate is known to accelerate the rate of spontaneous decarboxylation of 2-aminoisobutyrate by a factor of 10^{10} [12]

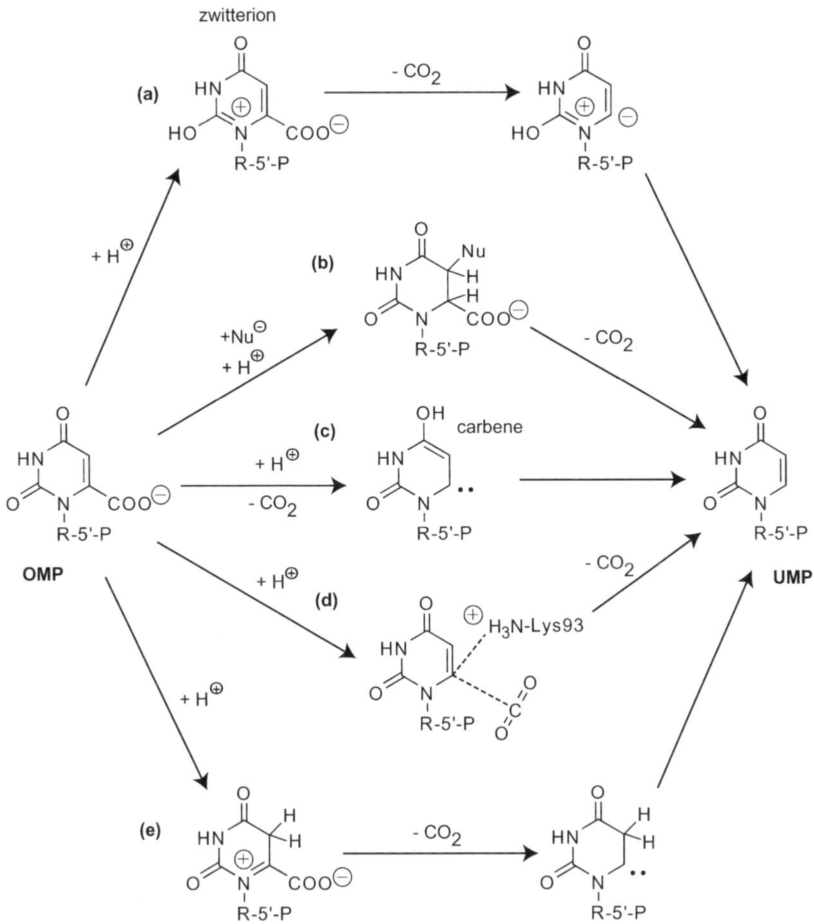

Fig. 3 Potential mechanisms of enzymatic decarboxylation as proposed by *a* Beak and Siegel [14]; *b* Silverman and Groziak [22]; *c* Lee and Houk [24]; *d* Appleby, Kinsland, Begley and Ealick [26]; and *e* Lee, Chong, Chodera, and Kollman [30]

tor (Fig. 2) resembles the structure of Beak and Siegel's 1,3-dimethylorotate zwitterion (Fig. 3a). This study also revealed the enzyme's preference for binding anionic ligands, further suggesting the presence of a positively charged active site residue. A subsequent search for this acidic residue resulted in the identification of Lys-93 of the yeast enzyme [16]. Mutation of Lys-93 to cysteine reduced activity by more than 10^7-fold, leading Smiley and associates to propose that Lys-93 functions to protonate the O-2 atom of OMP, thus promoting formation of the zwitterion. ^{13}C and solvent isotope effects on the enzymatic reaction were found to be consistent with a mechanism that included a proton-sensitive step preceding decarboxylation [17, 18].

To test the validity of the zwitterionic mechanism, Rishavy and Cleland conducted nitrogen isotope studies on the enzymatic decarboxylation of OMP [19]. Previous studies using viscogen effects and ^{13}C kinetic isotope measurements had demonstrated that C–C bond cleavage was rate-limiting for the enzyme reaction [17, 20]. Rishavy and Cleland detected no change in bond order at N-1 of substrate OMP during decarboxylation, suggesting that formation of a nitrogen ylide is not involved in the chemical mechanism of the enzymatic reaction. These results were consistent with later studies by Lee and coworkers on the non-enzymatic decarboxylation of 1,3-dimethylorotate, which indicate that decarboxylation occurs via formation of a neutral carbene rather than via formation of a ylide [21].

An alternate mechanism for enzymatic decarboxylation has been proposed by Silverman and Groziak [22]. By analogy to the reaction catalyzed by thymidylate synthase, these authors suggested that decarboxylation proceeds via addition of an active site nucleophile across the C5–C6 double bond of substrate OMP (Fig. 3b). Silverman and Groziak suggested that Michael addition at the C-5 position of the pyrimidine ring results in the formation of a covalent enzyme-substrate intermediate that can undergo an acid-base catalyzed elimination reaction to yield UMP and CO_2. The observation that a number of orotic acid analogs are susceptible to nucleophilic attack at C-5 lent support to this proposal. Efforts to confirm the addition mechanism proved unsuccessful when Acheson and coworkers failed to observe a change in hybrization at the C-5 position of bound ^{13}C-labeled BMP, and when no significant secondary isotope effect was observed upon decarboxylation of deuterium-labeled OMP [23]. These findings point to little or no change in geometry at C-5 in the transition state for decarboxylation, and cast doubt upon the likelihood of the Silverman and Groziak mechanism.

Lee and Houk calculated gas phase proton affinities of orotate and deprotonated uracil, which suggest that O-4 rather than O-2 is the favorable site of protonation for substrate OMP [24]. On the basis of these findings, Lee and Houk proposed a carbene-based mechanism that involves protonation at O-4 by either an active-site acidic residue or a site-bound water molecule (Fig. 3c) [24, 25]. In this mechanism, the formation of a neutral carbene at C-6 is stabilized by an active site environment that displays a low dielectric constant. The recent determination of the crystal structures of ODCase (see below) questions the plausibility of this mechanism. These structures reveal a highly charged active site, one that might be poorly suited for stabilization of an uncharged carbene. The structures also demonstrate the lack of an acidic residue near the O-4 atoms of bound ligands.

Based on the crystal structure of the *Bacillus subtilis* enzyme in complex with product UMP, Appleby and associates have proposed a novel, electrophilic substitution mechanism for the ODCase reaction (Fig. 3d) [26]. Instead of requiring the formation of a discrete carbanion in the transition state for decarboxylation, this mechanism postulates that Lys-93 protonates the fragmenting C–C bond once that bond becomes sufficiently basic. Several reports of electrophilic substitution reactions of this type have been reported in the literature [27, 28], although no evidence exists supporting this type of mechanism in an enzyme-catalyzed reaction. Recently, the acidity of the C-6 proton of 1,3-dimethyluracil was estimated to be 34, demonstrating the difficulty with which the C-6 carbanion is formed [29]. The spontaneous rate of cleavage of the C–H bond of dimethyluracil is 10-fold slower than the rate of spontaneous decarboxylation of 1-methylorotate, suggesting that ODCase may avoid formation of a discrete carbanionic species during decarboxylation.

Recently, Kollman and associates have suggested that enzymatic decarboxylation might proceed via direct protonation of substrate OMP at the C-5 position by Lys-93 of the yeast enzyme (Fig. 3e) [30]. This mechanism was suggested as a result of molecular dynamics simulations of the *Methanobacterium thermoautotrophicum* enzyme, which indicate that the side chain of Lys-93 might be positioned near the C-5 atom of OMP in the Michaelis complex. Similar to the Silverman and Groziak proposal, this mechanism is expected to produce a change in hybridization at C-5 during the course of the reaction. As noted above, experimental evidence suggests that rehybridization at C-5 does not occur during binding or catalysis.

3 Structure of Yeast ODCase in the Presence and Absence of a Potential Transition State Analog

Recently, the crystal structures of ODCase from yeast and from three different bacterial sources have been reported [26, 31, 32, 33]. Although a detailed comparison of the available crystal structures is beyond the scope of this review, the structure of the yeast enzyme both free and in complex with the postulated transition state analog, BMP, will be briefly discussed. For a detailed discussion of the currently available crystal structures, readers are referred to reference [34] and the article by Wu and colleagues in this issue.

The crystal structure of yeast ODCase reveals a $(\beta/\alpha)_8$-barrel fold similar to that first described for the glycolytic enzyme triosephosphate isomerase [35]. The active site of each monomer is located at the C-terminal end of the barrel and is composed of residues from both subunits of the catalytically active dimer. Similar to many $(\beta/\alpha)_8$-barrel enzymes that act on phosphorylated substrates, ODCase contains a flexible loop (residues 207–218 of the yeast enzyme) capable of forming hydrogen bonds with the phosphoryl group of bound ligands. In the absence of ligand, this loop is disordered. A second flexible loop, comprised of residues 151–165, is also observed in the crystal structure of the yeast enzyme. This loop contains residues that make both van der Waals and hydrogen bonding contacts with the pyrimidine ring

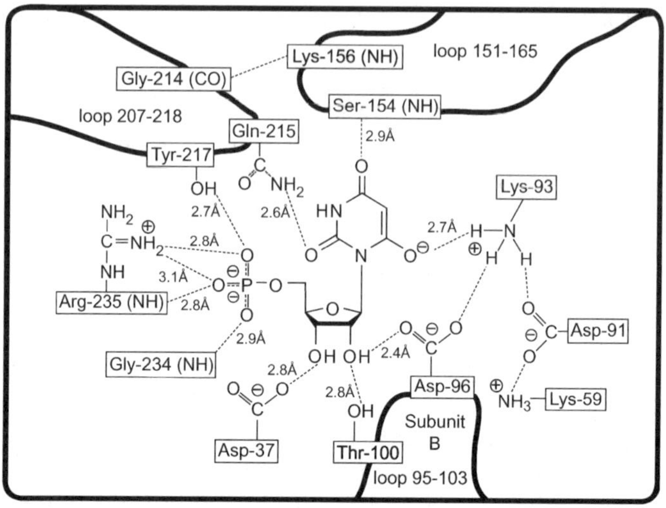

Fig. 4 Schematic representation of the yeast ODCase active site, showing contacts formed between the enzyme and bound 6-hydroxyuridine 5′-phosphate (BMP). Hydrogen bonding distances were measured between electronegative atoms

of bound BMP. Upon ligand binding loops 207–218 and 151–165 move across the active site to form a closely packed, closed conformation (Fig. 4). As a result of these combined loop movements, bound ligands are sequestered from bulk solvent water.

Interactions between yeast ODCase and BMP can be divided into three groups—those involving enzyme contacts with the phosphoryl group, ribosyl group, and pyrimidine ring of the inhibitor (Fig. 4). Binding of BMP promotes an ordering of residues in loop 207–218, allowing the formation of hydrogen bonds between the phosphoryl group and the side chains of Tyr-217 and Arg-235 [31]. Together with the backbone amide nitrogen of Gly-234, these residues form a total of five hydrogen bonds with the phosphoryl group of BMP. Three different active site residues are involved in hydrogen bonding contacts with the ribose hydroxyl groups of bound inhibitor. The 3′-OH group of BMP interacts with the side chain of Asp-37, and the 2′-OH group forms contacts with Thr-100 and Asp-96, two residues that are donated by the opposite subunit of the dimer. Somewhat surprisingly, the pyrimidine ring of bound BMP forms a rather limited number of interactions with active site residues. A single hydrogen bond is formed between the amide moiety of Gln-215 and O-2 of BMP, and the backbone amide nitrogen of Ser-154 forms a hydrogen bond with O-4. The sparsity of interactions with the pyrimidine ring offers one explanation for why purine analogs can act as effective competitive inhibitors of ODCase (Fig. 2). In total, 11 hydrogen bonds are observed between active site residues and BMP.

Perhaps the most intriguing feature of the ODCase active site is the presence of a quartet of charged residues near the site where the substrate's reactive carboxyl group is expected to bind. Composed of Lys-59, Asp-91, Lys-93, and Asp-96 of the opposite subunit, this network of alternating charges is evolutionarily conserved in all known ODCase sequences [36]. Prior to the determination of the ODCase crystal structure, Lys-93 had been shown to play an important role in enzyme-catalyzed decarboxylation [16]. In the structure of the yeast enzyme–BMP complex, this positively charged residue can be observed participating in an ionic interaction with the negatively charged O-6 atom of BMP. This favorable contact presumably foreshadows the interaction of Lys-93 with the developing C-6 carbanion formed during decarboxylation of OMP. The position of the Lys-93 side chain appears to be tightly constrained in the BMP complex by contacts with two neighboring aspartate residues (Asp-91 and Asp-96). Unlike the loop residues that contact the phosphoryl group and pyrimidine ring, the conformation of the four charged residues appears rigid, changing little between the apo and BMP liganded structures.

4 Contribution of Enzyme–Phosphoryl Contacts to Catalysis by ODCase

As described above, the structure of the complex between yeast ODCase and BMP reveals hydrogen bonding contacts between the phosphoryl group of this inhibitor and the side chains of Arg-235 and Tyr-217. To define the contributions of Arg-235 and Tyr-217 to transition state binding affinity, each of these residues was replaced with alanine [37]. Removal of Arg-235 was found to reduce k_{cat}/K_m by a factor of 3,000, whereas removal of Tyr-217 decreased k_{cat}/K_m by 7,500-fold. A mutant enzyme in which both Arg-235 and Tyr-217 were replaced with alanine demonstrated a decrease in activity of more than 3×10^7-fold, corresponding to a loss in transition state binding energy of more than 10 kcal/mol. The magnitude of the catalytic effects produced by eliminating the hydrogen bonds mediated by Arg-235 and Tyr-217 are particularly interesting considering the distance of these residues from the site of chemical transformation of substrate OMP.

The contribution of the phosphoryl group to catalysis was also assessed by evaluating the effects on ligand binding affinity of removing the phosphoryl group from a number of competitive inhibitors (Table 1) [37, 38]. These data suggest that the contribution of the phosphoryl group to ligand binding is relatively modest for weak binding inhibitors such as product UMP (120-fold), but can become very large as the affinity of the ligand increases. For example, the contribution of the phosphoryl moiety to the binding affinity of BMP, the most potent inhibitor of ODCase, exceeds 10^6-fold. This represents more than half of the 15 kcal/mol of binding energy realized

Table 1 Ligand dissociation constants for yeast ODCase

	K_i base (M)	K_i nucleoside (M)	K_i nucleotide (M)	K_i base/ K_i side	K_i side/ K_i tide
Uracil	1.6×10^{-2}	2.3×10^{-2}	2.0×10^{-4}	0.7	120
Orotate	9.5×10^{-3}	1.4×10^{-3}	7.0×10^{-7a}	6.8	2,000
Xanthine	$>6.2\times10^{-4}$	5.3×10^{-4}	4.1×10^{-7}	>1.2	1,300
6-Azauracil	5.9×10^{-3}	1.2×10^{-2}	6.4×10^{-8}	0.5	190,000
Barbiturate	6.0×10^{-6}	9.3×10^{-6}	8.8×10^{-12}	0.6	1,100,000

The contribution of the ribofuranosyl group for each ligand is calculated from the ratio of the dissociation constant of the base compared to the dissociation constant of the corresponding nucleoside. Similarly, the contribution of the phosphoryl group is calculated from the ratio of the dissociation constant of the nucleoside compared to the dissociation constant of the corresponding nucleotide

in the yeast ODCase–BMP complex. A comparison of the structure of the yeast ODCase–BMP complex with that of the *E. coli* ODCase–UMP complex offers no structural explanation for the differential contribution of the phosphoryl group to the binding affinity of these two ligands. Effects of similar magnitude have been observed with nucleoside and nucleotide inhibitors of the bifunctional human enzyme, and suggest that the charged phosphoryl group is a critical determinant of ligand binding affinity [38].

The effects of the R235A and Y217A mutations upon ligand binding affinity were found to resemble the effects, noted above, for ligands lacking a phosphoryl moiety (Table 2) [37]. Individual deletion of the Arg-235 or Tyr-217 side chain produced only slight decreases in binding affinity for the weakly bound UMP. Removal of these same side chains, however, produces pronounced decreases ($>10^3$-fold) in the binding constant of 6-azaUMP, a ligand that binds the wild-type enzyme more tightly than does UMP. These observations provide compelling evidence that contacts between the enzyme

Table 2 Decrease in ligand binding affinity resulting from mutagenesis of active site residues to alanine

Ligand	K_i (Y217A)/K_i (wt)	K_i (R235A)/K_i (wt)	K_i (K59A)/K_i (wt)
UMP	17	9	1.7
OMP	130	70	910
6-AzaUMP	2,200	1,400	1,100
BMP	2,000	2,600	11,000
Transition state	3,000	7,500	120,000

The relative contribution of each interaction was calculated from the ratio of the ligand dissociation constant determined with the mutant enzyme to that determined with wild-type yeast ODCase

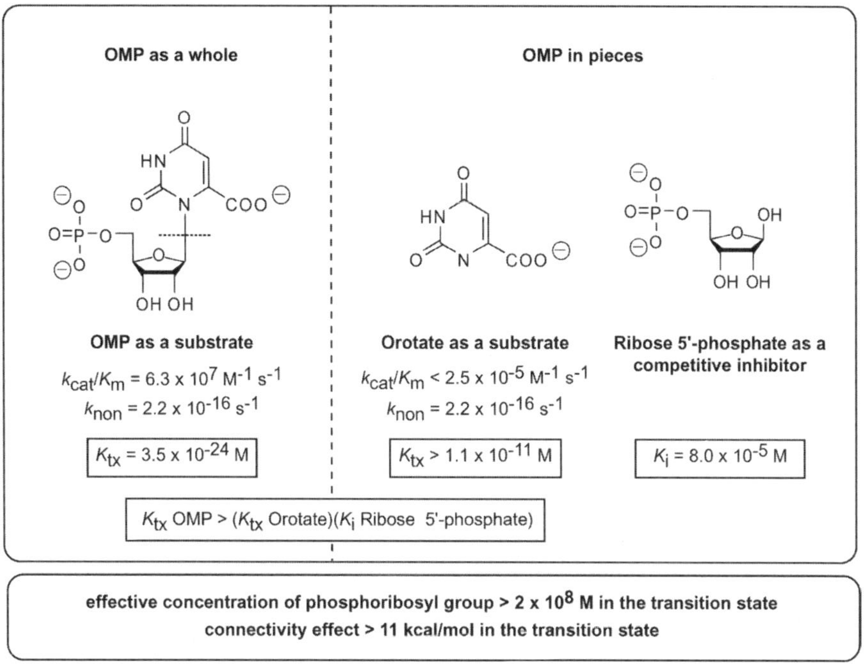

Fig. 5 Comparison of the binding affinities of yeast ODCase for OMP in the transition state for decarboxylation, orotate in the transition state for decarboxylation, and ribose 5′-phosphate acting as a competitive inhibitor. The loss of transition state binding energy obtained by cutting substrate OMP at the glycosidic bond totals more than 11 kcal/mol

and the phosphoryl group become increasingly important as the substrate passes from the ground state to the transition state.

To examine the role of the entire phosphoribosyl group in the enzymatic decarboxylation of OMP, orotic acid was tested as a substrate for yeast ODCase [37]. Carrier-free orotic acid labeled with ^{14}C at the carboxyl substituent was incubated with high concentrations of enzyme in an effort to observe slow decarboxylation of this truncated substrate. Orotic acid was found to decarboxylate at a rate more than 10^{12}-fold slower than OMP, and 1 M inorganic phosphate was unable to rescue this activity. Comparison of the affinity of the enzyme for OMP in the transition state, orotic acid in the transition state, and ribose 5′-phosphate as a competitive inhibitor, indicates that the effective concentration of the phosphoribosyl group in the transition state for decarboxylation exceeds 10^8 M (Fig. 5). Thus, a significant portion of transition state affinity (>11 kcal/mol) can be attributed to the covalent linkage between orotic acid and ribose 5′-phosphate. A similar contribution of a remote functional group to enzyme catalysis was recently described for the isomerization of glyceraldehyde 3-phosphate by triose-phosphate isomerase (TIM) [39]. In this case, the phosphoryl group of the substrate was found to contribute 14 kcal/mol of intrinsic binding energy in the transition state for enolization. For both the TIM and ODCase-catalyzed reactions, the substrate's phosphoryl group appears to be an important feature by which the enzyme distinguishes the altered substrate in the transition state from the ground state.

The $>10^{12}$-fold difference in transition state binding affinity between OMP and its truncated analog is surprising considering the relatively modest difference in ground state binding affinity between orotic acid (K_m=9.3×10^{-3} M) and OMP (K_m=7.0×10^{-7} M). There is no structural or experimental evidence to suggest direct participation of the substrate's phosphoryl group in catalysis. The phosphoryl group of bound BMP is positioned more than 4.5 Å from the O-2 atom of the pyrimidine ring, and the intervening distance is filled by the side chain of Gln-215 (Fig. 4). Rather than participating directly in the decarboxylation reaction, the phosphoryl group seemingly acts as a repository for a significant portion of the substrate's intrinsic binding energy. The surplus of binding energy residing in the phosphoryl group is only fully realized in the transition state for decarboxylation, where it supplies part of the differential binding affinity upon which catalysis depends.

5
Contribution of Enzyme–Ribofuranosyl Contacts to Ground State and Transition State Affinity

Given the importance of the phosphoryl group noted above, the possibility existed that enzyme–ribofuranosyl contacts might play an equally important role in catalysis. The structure of the yeast enzyme complexed to BMP revealed interactions between Thr-100 and Asp-37 and the 2′-OH and 3′-OH groups of the ribose ring, respectively. To investigate the contribution of these ribofuranosyl contacts to both ground state and transition state stabilization, Thr-100 and Asp-37 were individually replaced with alanine [40]. Removal of Thr-100 reduced k_{cat}/K_m by 60-fold, while removal of the side chain of Asp-37 reduced k_{cat}/K_m by 300-fold. Ground state binding affinity of substrate OMP was reduced by less than 10-fold for both of these mutant enzymes, indicating a modest role for these contacts in determining ground state affinity (Table 3).

In a previous study, the 2′-OH group of substrate OMP had been postulated to form a hydrogen bond to the critically important Lys-93 side chain in the ground state ES complex [32]. In addition, the 2′-OH group serves as the sole site of contact between bound ligand and residues donated by the opposite subunit of the ODCase dimer (Fig. 4). To evaluate the catalytic importance of this functional group, 2′-deoxyOMP was synthesized and the ability of ODCase to catalyze the decarboxylation of this wounded substrate was as-

Table 3 Summary of the kinetic properties of wild-type and mutant yeast ODCases

Enzyme	k_{cat} (sec^{-1})	K_m (M)	k_{cat}/K_m (M^{-1} sec^{-1})	ΔΔG (kcal/mol)
Wild-type	44	7.0×10^{-7}	6.3×10^{7}	
Q215A	41	2.5×10^{-6}	1.6×10^{7}	0.8
T100A	7.2	6.7×10^{-6}	1.1×10^{6}	2.4
D37A	0.9	4.2×10^{-6}	2.1×10^{5}	3.4
Y217A	2.0	9.4×10^{-5}	2.1×10^{4}	4.7
R235A	0.42	4.9×10^{-5}	8.6×10^{3}	5.3
K59A	0.34	6.4×10^{-4}	5.3×10^{2}	6.9
D91A	n.d.	n.d.	n.d.	>7.0
K93A	n.d.	n.d.	n.d.	>7.0
D96A	$<2.3\times10^{-4}$	8.0×10^{-6}	<29	>8.6

n.d. Not determined due to lack of measurable activity [41]

sessed [40]. 2′-deoxyOMP displays a K_m value that is 8-fold larger than OMP, and a value for k_{cat} that is reduced by more than 200-fold. In agreement with a limited contribution of the ribosyl contacts to ground state binding, deletion of the 2′-OH group from two competitive inhibitors, UMP and ribose 5′-phosphate, produced little decrease in their observed binding affinities.

The contribution of the ribofuranosyl group as a whole was measured by comparing the affinities of a variety of related base and nucleoside analogs for yeast ODCase (Table 1). The addition of the ribofuranosyl moiety to the substrate analog orotic acid, the product analog uracil, and three different competitive inhibitors failed to increase the binding affinity of these compounds. Similar to the catalytic consequences of the D37A and T100A mutations, these comparisons indicate that hydrogen bonds involving the ribose hydroxyl groups are much less important to ground state and transition state binding than are contacts with the phosphoryl moiety. As observed with enzyme–phosphoryl interactions, however, hydrogen bonds involving Asp-37 and Thr-100 do appear to become increasingly important as the substrate passes from the ground state to the transition state. Therefore, the contribution of the 2′-OH and 3′-OH groups to transition state binding affinity is likely magnified when contacts involving these substituents act in concert with other enzyme–ligand interactions.

6
Importance of the Charged Network to Catalysis by ODCase

The ODCase crystal structures reveal a novel array of charged residues located in a region of the active site where the substrate's reactive carboxylate is expected to bind. In an attempt to elucidate the catalytic function of each of these residues, Lys-59, Asp-91, Lys-93, and Asp-96 of the yeast enzyme were replaced with alanine [41]. Mutagenesis of Asp-96 reduced activity below the limits of detection, but did not impair the ability of this enzyme to bind OMP to a significant degree. The dissociation constant of the D96A-OMP complex determined by isothermal titration calorimetry (ITC) was 8.0×10^{-6} M, a value 10-fold higher than the observed K_m of the wild-type enzyme. Similar to the D96A mutation, substitution of Lys-59 with alanine had a profound effect upon catalysis, decreasing k_{cat}/K_m by 10^5-fold. Specifically, k_{cat} was reduced 130-fold, and K_m was increased by nearly three orders of magnitude (Table 3). Effects of this magnitude can be explained, in part, by the fact that the K59A mutant active site contains an unbalanced negative charge. This net negative charge reduces the enzyme's affinity for the negatively charged substrate in the ground state and, to a greater degree, for the negatively charged transition state.

Replacement of either Asp-91 or Lys-93 with alanine reduced activity by more than 5 orders of magnitude, and produced enzymes that were incapable of binding substrate OMP at a level detectable by ITC. These findings are consistent with recent denaturation studies of *M. thermoautotrophicum* mutant enzymes, which suggest that preservation of the charged network is critical to the stability of the enzyme [42]. Together, the results of mutagenizing Lys-59, Asp-91, Lys-93, and Asp-96 indicate that removal of even a single member of the charged active site network severely impairs the catalytic capabilities of ODCase.

The role of Lys-93 in stabilizing the development of negative charge at the C-6 position of OMP in the transition state for decarboxylation appears to be clear. The catalytic function of Asp-91, however, is more ambiguous. Modeling OMP into the binding site of 6-azaUMP from the *M. thermoautotrophicum* crystal structure, Wu and associates found significant electrostatic and steric repulsion between the negatively charged side chain of Asp-91 and the substrate's reactive carboxylate [32]. To understand the role this repulsion might play in enzymatic decarboxylation, the authors performed combined quantum mechanical and molecular mechanics calculations of the enzymatic reaction. The results of this study led the authors to hypothesize that a portion of the binding energy supplied by the phosphoribosyl group of OMP is used to destabilize the reactive carboxyl group in the ground state ES complex by positioning it near Asp-91. The resulting electrostatic stress was expected to supply a significant portion of the activation energy required to overcome the barrier to decarboxylation.

In an effort to validate these findings, Warshel and coworkers carried out similar calculations on the enzyme-catalyzed reaction [43, 44]. Unlike the calculations of Wu and colleagues, these authors were able to reproduce the catalytic effect of the enzyme without invoking ground state destabilization. Using a modified reactant pair composed of the orotate anion and the positively charged lysine side chain, Warshel and coworkers found evidence for substantial stabilization of the substrate in the ground state complex. Instead of observing ground state destablization between the substrate and the enzyme, electrostatic destabilization was observed between neighboring aspartate residues (Asp-91 and Asp-96) within the enzyme itself[2]. The authors suggested that the price of this destabilization is paid for by the energetics of protein folding, rather than by utilization of the large intrinsic binding free energy supplied by the substrate's phosphoribosyl group. This study concluded that ODCase functions by preferentially stabilizing the transition state, rather than by inducing destabilization of the ground state.

In a more recent molecular dynamics study, Bruice and associates also found evidence supporting stabilization of substrate OMP in the ground state complex [45]. In these studies, the active site loop that contacts the phosphoribosyl moiety was shown to be quite flexible, a suggestion that precludes its ability to force the substrate's reactive carboxylate into an unfavorable environment. The structure of the ground state OMP complex observed in these simulations reproduces the structure of OMP in aqueous solution, and the reactive carboxylate of OMP does not approach the side chain carboxylate of Asp-91 in the ES complex. This study attributes the large catalytic contribution of the substrate's phosphoribosyl group to correct positioning of the carboxyl group with respect to Lys-93, rather than a direct utilization of the phosphoribosyl's binding energy for ground state destabilization.

At first glance, it is difficult to envision how significant destabilization might occur between the substrate's reactive carboxylate group and the side chain of Asp-91. Either carboxyl group could, in principle, abstract a proton from bulk solvent water if it were faced with significant electrostatic repulsions. UMP, which lacks the carboxyl group of substrate OMP, binds to the wild-type enzyme much less tightly than does the stressed substrate (Fig. 1). The opposite situation might be expected if the ground state OMP complex was destabilized with respect to the product UMP complex. Similarly, removal of the repulsive Asp-91 side chain via mutagenesis fails to result in an enzyme that displays an increased affinity for OMP. Nevertheless, the profound cooperativity of enzyme–ligand contacts, as revealed by experiments with the phosphoryl mutants and phosphoryl deficient ligands, may mask the destabilizing effects of Asp-91 in these cases. Interestingly, a recently described structure of the D91A-K93A double mutant in complex with OMP shows the reactive carboxylate moiety bent out of the plane of the pyrimidine ring, in a conformation distinct from those previously observed for orotate in aqueous solution [42]. While it seems likely that some degree of distortion of the reactive carboxyl group does occur, the extent to which this

[2] This finding is consistent with the experimental observation that mutagenesis of Asp-91 or Lys-93 decreases the stability of the mutant enzyme [41, 42]

higher energy conformation mirrors the structure of substrate OMP in the ground state ES complex remains to be resolved.

7 Summary and Future Prospects

The results of the mutagenesis studies summarized above, in combination with the structures of numerous ODCase—inhibitor complexes, have provided valuable insight into the forces at work within the ODCase active site during decarboxylation. Contacts involving both the phosphoryl group of substrate OMP and the active site charged network have been identified as important determinants of transition state binding affinity. Mutagenesis demonstrates that the total amount of energy available to bind the altered substrate in the transition state, attained by summing the contributions of individual active site interactions, exceeds 45 kcal/mol (Fig. 6). This far surpasses the binding energy (~31 kcal/mol) needed to explain the high level of catalytic proficiency displayed by ODCase. The phosphoribosyl group of substrate OMP appears to anchor the substrate in the ODCase active site, thereby supplying a vast quantity of the intrinsic binding energy necessary for decarboxylation to occur. This group also appears to be a major feature

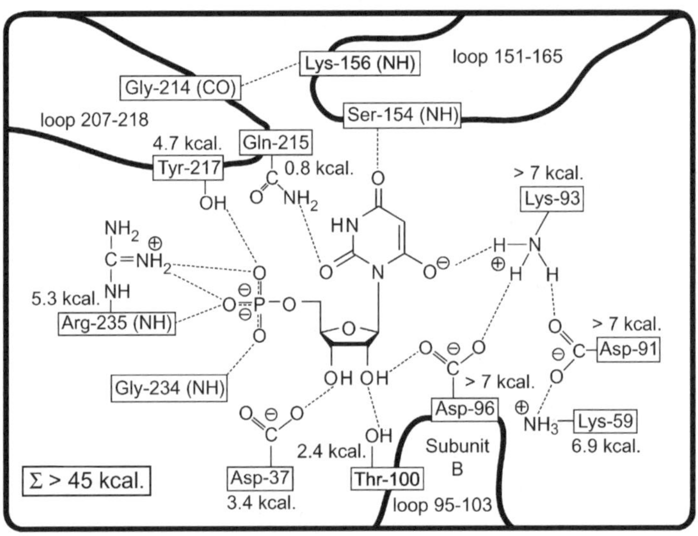

Fig. 6 Schematic representation of the yeast ODCase active site, depicting the effects on transition state binding affinity of replacing individual active site residues with alanine. Free energy contributions ($\Delta\Delta G$) of individual interactions were calculated from the decrease in catalytic efficiency (k_{cat}/K_m) produced upon mutation to alanine. The sum of individual mutations totals more than 45 kcal/mol of binding free energy

by which the enzyme distinguishes the substrate in the ground state from the altered substrate in the transition state.

Further mechanistic descriptions of the ODCase reaction await the elucidation of ODCase–ligand structures that more closely mimic the structure of the Michaelis complex. In this regard, two particularly pertinent complexes whose structures remain undetermined include ODCase bound to 5,6-dihydroorotidine 5′-phosphate, and 6-thiocarboxamidouridine 5′-phosphate (Fig. 2). Both of these ligands contain groups at the C-6 position that closely resemble either the charge or geometry of the carboxylate group of substrate OMP. Thus, the structure of ODCase in complex with either ligand promises to reveal active site conformations that are specific to the ground state ES complex. In addition, 5,6-dihydroOMP appears to approximate the structure of an intermediate formed during the electrophilic substitution mechanism proposed by Appleby and associates (Fig. 3d). Although this ligand does not appear to function as a substrate for decarboxylation, 5,6-dihydroOMP does act as a competitive inhibitor, binding to the yeast enzyme with an affinity near that of OMP [46].

The structure of ODCase complexed to 6-thiocarboxamidoUMP is of interest because this ligand binds 10^5-fold more tightly to the yeast enzyme than does its oxygen containing counterpart 6-carboxamidoUMP [47]. The hydrophobicities of these two molecules are not expected to differ greatly, however the difference in polarizability between sulfur and oxygen atoms has been well documented in the literature [48, 49]. To understand the large difference in affinity between the sulfur- and oxygen-containing inhibitors, it is helpful to examine the electronic differences between the oxygen- and sulfur-containing functional groups. Figure 7 shows the three main resonance hybrid structures for the side chains of 6-carboxamidoUMP and 6-thiocarboxamidoUMP. For the sulfur derivative, species II and III are expected to contribute significantly to the overall structure of the thio compound, since the orbital overlap between carbon and the bulky sulfur atom

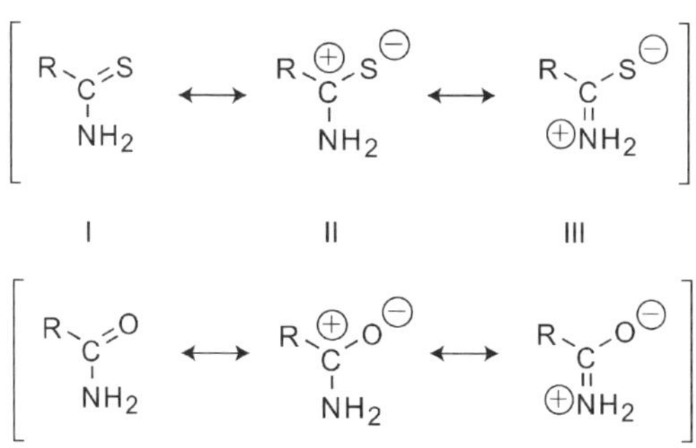

Fig. 7 Major resonance hybrid structures of 6-thiocarboxamidoUMP and 6-carboxamidoUMP

in the C=S bond is inefficient [48]. In contrast, the analogous species for the oxy derivative should not contribute as significantly to the overall structure, since orbital overlap between carbon and the smaller oxygen atom is quite favorable. Thus, thiocarboxamidoUMP is expected to contain more localized anionic character at the C-6 position than is carboxamidoUMP. The large difference in binding affinity produced by this minor difference in charge distribution may reflect the ability of the enzyme to distinguish small changes in electron density at the C-6 position during decarboxylation.

8 Concluding Remarks

To achieve catalysis, an enzyme must bind the altered substrate in the transition state orders of magnitude more tightly than it binds the substrate in the ground state. For the ODCase reaction, this task is complicated by the fact that both the substrate and the transition state contain negative charge at the site of chemical transformation. How does the active site architecture of ODCase provide for the level of discrimination observed between the binding of the substrate in the ground state (K_s=10^{-7} M) and the binding of the altered substrate in the transition state (K_{tx}=10^{-24} M)? To achieve preferential binding of the transition state, the enzyme must prevent favorable interactions in the ground state complex [50]. In particular, the enzyme must prevent formation of a hydrogen bond between Lys-93 and the reactive carboxyl group of substrate OMP. Upon substrate binding Lys-93 can, in principle, interact with either the carboxyl group of OMP or with a different hydrogen bonding partner within the active site. Which partner Lys-93 chooses depends upon which interaction is more energetically favorable. In the apo and BMP-liganded yeast ODCase crystal structures, Lys-93 is positioned between two negatively charged aspartate residues (Asp-91 and Asp-96). These interactions probably raise the pK_a value of Lys-93 above that of a lysine side chain in aqueous solutions (~10.5). The elevation of this pK_a value likely minimizes the strength of contacts between the slightly basic carboxylate group of OMP (pK_a~0.4) and the slightly acidic lysine side chain. Given the weakness of this potential interaction, it seems probable that Lys-93 moves away from the reactive carboxylate group in the ground state to form a more favorable hydrogen bond with another enzyme functional group. Upon approach to the highly charged transition state, however, the lysine side chain sacrifices its ground state, enzyme-based contact for a more favorable interaction with the developing carbanion at C-6. At this same moment, the strength of enzyme contacts with the substrate's phosphoryl group are increasing in intensity. The realization of binding energy associated with the newly formed carbanion–lysine contact, in combination with a tightening of interactions involving the substrate's phosphoryl group, produces the differential binding affinity upon which catalysis relies.

Acknowledgements The author wishes to thank Professors Richard Wolfenden, W. Wallace Cleland, and Ronald Raines for valuable discussions and continued support. The author also thanks Bryan Smith and Betsy Kersteen for editorial comments. BGM is a Damon Runyon Fellow Supported by the Damon Runyon Cancer Research Foundation (DRG-1702).

References

1. Jones ME (1980) Annu Rev Biochem 49:253
2. Lieberman I, Kornberg A, Simms ES (1954) J Am Chem Soc 76:2844
3. Wright LD, Miller CS, Skeggs HR, Huff JW, Weed LL, Wilson DW (1951) J Am Chem Soc 73:1898
4. Suttle DP, Becroft DMO, Webster DR (1989) In: Schriver CR, Beaudet AL, Sly WS, Valle D (eds) The metabolic basis of inherited disease. McGraw-Hill, New York, p 1095
5. Handschumacher RE, Calabresi P, Welch AD, Bono V, Fallon H, Frei E (1962) Cancer Chemother Rep 21:1
6. Sweeney MJ, Davis FA, Gutwosk GE, Hamill RL, Hoffman DH, Poore GA (1973) Cancer Res 33:2619
7. Radzicka A, Wolfenden R (1995) Science 267:90
8. Snider MJ, Wolfenden R (2000) J Am Chem Soc 122:11507
9. Shostak K, Jones ME (1992) Biochemistry 31:12155
10. Cui W, DeWitt JG, Miller SM, Wu W (1999) Biochem Biophys Res Comm 259:133
11. Miller BG, Smiley JA, Short SA, Wolfenden R (1999) J Biol Chem 274:23841
12. Zabinski RF, Toney MD (2001) J Am Chem Soc 123:193
13. Beak P, Siegel B (1973) J Am Chem Soc 95:7919
14. Beak P, Siegel B (1976) J Am Chem Soc 98:3601
15. Levine HL, Brody RS, Westheimer FH (1980) Biochemistry 19:4993
16. Smiley JA, Jones, ME (1992) Biochemistry 31:12162
17. Smiley JA, Paneth P, O'Leary MH, Bell JB, Jones ME (1991) Biochemistry 30:6216
18. Ehrlich, JI, Hwang C-C, Cook PF, Blanchard JS (1999) J Am Chem Soc 121:6966
19. Rishavy MA, Cleland WW (2000) Biochemistry 39:4569
20. Miller BG, Snider MJ, Short SA, Wolfenden R (2000) Biochemistry 39:8113
21. Singleton DA, Merrigan SR, Kim BK, Beak P, Phillips LM, Lee JK (2000) J Am Chem Soc 122:3296
22. Silverman RB, Groziak MP (1982) J Am Chem Soc 104:6436
23. Acheson SA, Bell JB, Jones ME, Wolfenden R (1990) Biochemistry 29:3198
24. Lee JK, Houk KN (1997) Science 276:942
25. Houk KN, Lee JK, Tantillo DJ, Bahmanyar S, Hietbrink BN (2001) ChemBioChem 2:113
26. Appleby TC, Kinsland C, Begley T, Ealick SE (2000) Proc Natl Acad Sci USA 97:2005
27. Gawley RE, Low E, Zhang Q, Harris R (2000) J Am Chem Soc 122:3344
28. Futuko JM, Jensen FR (1983) Acc Chem Res 16:177
29. Sievers A, Wolfenden R (2002) J Am Chem Soc 124:13986
30. Lee T-S, Chong LT, Chodera JD, Kollman PA (2001) J Am Chem Soc 123:12837

31. Miller BG, Hassell AM, Wolfenden R, Milburn MV, Short SA (2000) Proc Natl Acad Sci USA 97:2011
32. Wu N, Mo Y, Gao J, Pai EF (2000) Proc Natl Acad Sci USA 97:2017
33. Harris P, Navarro-Poulsen JC, Jensen KF, Larsen S (2000) Biochemistry 39:4217
34. Begley TP, Appleby TC, Ealick SE (2000) Curr Opin Struct Biol 10:711
35. Banner DW, Bloomer AC, Petsko GA, Phillips DC, Pogson CI, Wilson IA, Corran PH, Furth AJ, Milman JD, Offord RE, Priddle JD, Waley SG (1975) Nature 255:609
36. Traut TW, Temple B (2000) J Biol Chem 275:28675
37. Miller BG, Snider MJ, Short SA, Wolfenden R (2000) Biochemistry 39:8113
38. Miller BG, Traut TW, Wolfenden R (1998) Bioorg Chem 26:283
39. Amyes TL, O'Donoghue AC, Richard J (2001) J Am Chem Soc 123:11325
40. Miller BG, Butterfoss GL, Short SA, Wolfenden R (2001) Biochemistry 40:6227
41. Miller BG, Snider MJ, Short SA, Wolfenden R (2001) J Biol Chem 276:15174
42. Wu N, Gillon W, Pai EF (2002) Biochemistry 41:4002
43. Warshel A, Strajbl M, Villa J, Florian J (2000) Biochemistry 39:14728
44. Warshel A, Florian J, Strajbl M, Villa J (2001) ChemBioChem 2:109
45. Hur S, Bruice TC (2002) Proc Natl Acad Sci USA 99:9668
46. Miller BG, Wolfenden R (2002) Annu Rev Biochem 71:847
47. Landesman PW (1982) PhD thesis, SUNY Buffalo
48. Idoux JP, Hwang PTR, Hancock CK (1973) J Org Chem 38:4239
49. Abboud JLM, Roussel C, Gentric E, Sraidi K, Lauransan J, Guiheneuf G, Kamlet M, Taft RW (1988) J Org Chem 53:1545
50. Schowen RL (1978) In: Gandor RD, Schowen RL (eds) Transition states of biochemical processes. Plenum, New York, p 78

Toc Curr Chem (2004) 238:63–78
DOI 10.1007/b94539

Survey of Enzymological Data on ODCase

Jeffrey A. Smiley (✉)

Department of Chemistry, Youngstown State University, Youngstown, OH 44555, USA
jasmiley@cc.ysu.edu

Abstract OMP decarboxylase (ODCase) has been the subject of enzymological, structural, and theoretical studies for several decades, yet its highly proficient rate of decarboxylation remains mechanistically unclear. This review summarizes the enzymological and model studies carried out with ODCase and its substrate analogs. The original mechanism involving protonation of O2 of the substrate OMP by an enzyme functional group has gained some experimental support, including the measurement of isotope effects, affinities of various inhibitors, and catalytic activity or inactivity toward alternate substrates. A lysine residue of the yeast enzyme, Lys93, invariant in all known sequences and situated within a region of the sequence that is highly similar throughout all sequences, has been proposed to be the active site proton donor. However, the crystal structures of ODCase in complex with various inhibitors show the 5,6 side of the pyrimidine of the inhibitors in proximity with the active site lysine side chain, with O2 turned away from the lysine. The enzymological data is reconciled with the structural data in this review with the proposal that the nearly symmetrical inhibitors bind in an orientation that is reversed from that of the substrate, with the glycosidic bond rotated roughly 180° from that projected for the substrate.

Keywords OMP decarboxylase · Isotope effects · Alternate substrate · Enzyme inhibition · Enzyme structure/function

Abbreviations

6-azaUMP	6-Azauridine-5′-monophosphate
BMP	1-(5′-Phospho-β-d-ribofuranosyl)barbituric acid
OMP	Orotidine-5′-monophosphate
2-thioOMP	2-Thioorotidine-5′-monophosphate
UMP	Uridine-5′-monophosphate

1 Introduction

OMP decarboxylase (ODCase) catalyzes the decarboxylation of OMP to UMP, a decarboxylation that must necessarily be mechanistically different from the groups of decarboxylations that occur throughout metabolism [1]. The structure of the substrate does not lend itself to decarboxylation mechanisms involving pyridoxal phosphate (typical of amino acid decarboxylases [2]), thiamine pyrophosphate (typical of α-keto acid decarboxylases [3]), or metal ions (typical of β-keto acid decarboxylases [4]); although the presence of Zn^{+2} ions has been detected in some preparations of ODCase [5, 6], the enzyme clearly does not require Zn^{+2} for catalytic activity [7].

Despite various mechanistic studies over several decades and the recent crystal structures of ODCases from four microbial species in the presence of various inhibitors, the mechanism by which ODCase catalyzes decarboxylation of OMP remains unclear. At least five mechanisms for the generation of a more reactive intermediate have been proposed on the basis of different experimental and theoretical approaches (Fig. 1). In this review, the enzymological and model chemistry experiments designed to illuminate the nature of the more reactive ODCase intermediate will be surveyed.

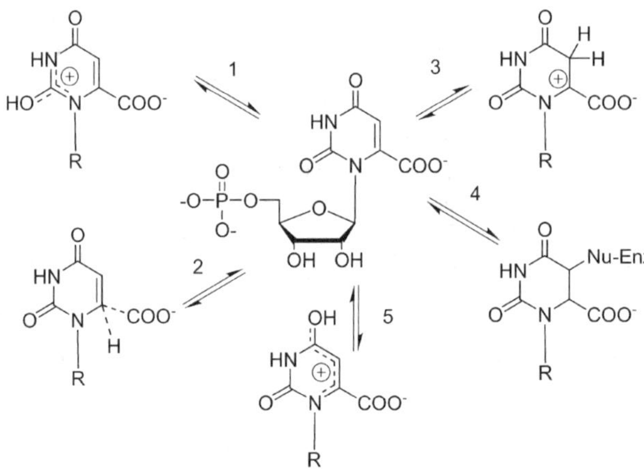

Fig. 1 Five proposed enzyme-catalyzed pre-decarboxylation reactions for the activation of OMP to a more reactive intermediate by ODCase. *1* O2 protonation mechanism, originally proposed in [8]. *2* Direct protonation of C6 during decarboxylation [9, 10, 11]. *3* Protonation at C5 [12]. *4* Nucleophilic attachment at C5 [13]. *5* Protonation at O4 [14]

2
The Uncatalyzed Decarboxylation of OMP Analogs

Beak and Siegel [8] presented a thorough study of the decarboxylation of 1,3-dimethylorotate and analogs in non-aqueous solvents. The decarboxylation of 1,3-dimethylorotic acid was found to be first order, and slightly faster with the anion 1,3-dimethylorotate than with the acid, indicating a decarboxylation of the carboxylate without general acid catalysis (Fig. 2).

By comparing the rates of decarboxylation of various analogs, the authors concluded that the mechanism of the uncatalyzed decarboxylation of 1,3-dimethylorotate likely involves formation of a dipole stabilized carbanion, with charges distributed around the resonance-stabilized heterocycle and an exocyclic oxygen bearing negative charge. Analogs of 1,3-dimethylorotate incapable of forming an analogous decarboxylated product generally had lower or unmeasurable rates of decarboxylation. A dimethoxypyrimidine carboxylate with a permanent positive charge distributed around the heterocycle underwent decarboxylation at a rate 10^7 times faster than 1,3-dimethylorotate, when corrected for temperature difference. The authors postulated that the role of ODCase might be to shift the equilibrium of OMP toward the formation of a protonated intermediate, resembling the permanently positive charged model compound, and suggested O2 as the likely site of protonation.

Radzicka and Wolfenden [15] extended the Beak and Siegel approach to include the decarboxylation of 1-methylorotate in water, which required reactions at approximately 200 °C under pressure. The decarboxylation in water is highly temperature-dependent, and the Arrhenius analysis yielded a rate constant of 3×10^{-16} s^{-1} when extrapolated to 25 °C. Using this comparison, ODCase, with k_{cat}=40 s^{-1}, must therefore accelerate the decarboxylation

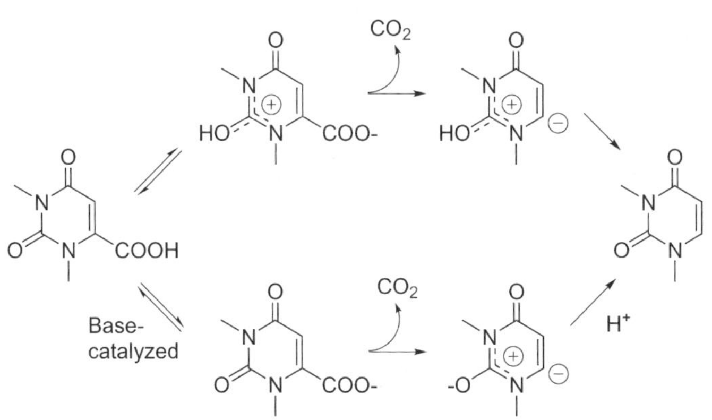

Fig. 2 Proposed mechanism for non-enzymatic decarboxylation of 1,3-dimethylorotic acid, in the absence (*upper route*) or presence (*lower route*) of added base. From [8]

reaction by 10^{17}-fold. Interestingly, the authors also observed no rate change between pH 3 and 11, indicating that the mechanism of decarboxylation in aqueous solution at neutral pH is independent of protonation, which is in accord with the lower path of Fig. 2, beginning with the carboxylate. However, protonation as a mechanistic step by ODCase could occur using a highly organized enzyme active site where the localized $[H^+]$ is much higher than that attainable in solution, which could resemble the upper path of Fig. 2 (or any H^+-dependent mechanism in Fig. 1) and represent a subtle departure from the reaction in aqueous solution.

3
Enhanced Affinity of Nucleotide Inhibitors with Anionic Nitrogenous Rings

The Westheimer group, noting the inhibition of ODCase by 6-azaUMP, synthesized a custom affinity column with 6-azaUMP appended to a stationary phase [16]. With this column, ODCase could be purified by a remarkable 3,200-fold in one purification step. With the availability of highly-purified enzyme, this group measured inhibition constants for UMP, 6-azaUMP and "barbituric acid ribonucleotide", or BMP [17].

The measured inhibition constants of UMP and 6-azaUMP are 4.6×10^{-4} M and 5.1×10^{-7} M at pH 6.0 [17]. Why would such structurally similar nucleotides have such different inhibition constants? The Westheimer group noted that UMP and 6-azaUMP differ primarily in the ionization constants of the nitrogenous rings: the uracil ring of UMP has a pK_a of 9.5, and the 6-azauracil ring of 6-azaUMP has a pK_a of 7.0. With the assumption that the anionic form of each inhibitor was the predominant form binding to the ODCase active site, the intrinsic dissociation constants of the anionic forms were calculated to be 1.5×10^{-7} M and 0.46×10^{-7} M, very similar values for nearly identical inhibitors. Nearly identical inhibition constants were later found for UMP and 2-thioUMP, when correcting for predominant inhibition by the anionic pyrimidine [18].

BMP has a lower pK_a of 4.5 and remains predominantly anionic at neutral pH levels. This inhibitor has a remarkably strong affinity for ODCase, with a near-stoichiometric binding constant of 9×10^{-12} M [17]. However, BMP possesses a structural twist: with the substituent ionized oxygen at C6, the pyrimidine ring, now nearly symmetrical, bears the negative charge on the C4-C6 side of the ring, unlike ionized UMP and 6-azaUMP with the charge delocalized on the C2-C4 side.

Suspecting that the preference of ODCase for anionic inhibitors might indicate the presence of a cationic active site amino acid residue, perhaps acting as a general acid in catalysis, Smiley and Jones sought to identify critical amino acid residues by site-directed mutagenesis, with particular attention to lysine residues [19]. By the late 1980s, quite a few ODCase sequences had been determined [20, 21], owing to the ease in isolation of this gene by microbial complementation. Lys93 of the yeast sequence was found to be present within a highly conserved region and was itself invariant in every single sequence, and its replacement with any of 15 other amino acids led to diminished activity [19]. The mutant ODCase K93C was studied in detail, and its activity was undetectable, with no more than 2×10^{-8} the activity of the wild type protein. Furthermore, the mutant ODCase had lowered affinity for 6-azaUMP. Chemical modification of K93C ODCase with 2-bromoethylamine—but not with 2-bromoethanol, 2-bromoacetamide, or (2-bromoethyl)trimethylammonium—led to partial restoration of activity, indicating a need at the active site for an amino group that none of the other functional groups could satisfy. A role for Lys93 as a proton donor in ODCase activity, and a role of attracting the anionic pyrimidine inhibitors, was proposed.

4
Isotope Effects in the ODCase Reaction

The Wolfenden and Jones groups addressed the possibility of an ODCase mechanism involving covalent attachment at C5 (mechanism 4, Fig. 1) [22], previously proposed on the basis of model chemistry studies [13]. Such a mechanism would be likely to display a kinetic isotope effect with [^{2}H5] OMP as the substrate, since this labeled substrate would involve an sp^2 to sp^3 hybridization change at C5 and a secondary kinetic isotope effect in the proposed pre-decarboxylation covalent step [23]. The measured isotope effect was essentially unity, 1.00±0.06, which would be diagnostic of a mechanism involving no significant chemistry at C5 except for the possibility that the enzyme was using a mechanism with a large forward fractionation factor, i.e., a "sticky" substrate that nearly always proceeded toward catalysis upon binding to the active site. A large forward fractionation factor would mask the observation of a secondary hydrogen isotope effect in this mechanism.

The Jones and O'Leary groups then measured the carbon isotope effects at the substrate carboxylate under a number of different conditions [24]. If ODCase were using a mechanism where the substrate is sticky, the carbon isotope effects would also be near unity. However, under some conditions, the observed isotope effect approached the expected value of the intrinsic isotope effect (the isotope effect on the individual C–C bond-breaking step). The ODCase reaction is thus necessarily one in which substrate binding is freely reversible, and only small contributions to the overall rate are made by pre-decarboxylation steps.

In their discussion, the authors used a mathematical analysis to address the C5 covalent mechanism using the available isotope effect data. The secondary hydrogen isotope effect was an effect on the kinetic constant V/K, indicated using conventional terminology [25, 26] as $^{D}(V/K)$, and is defined as follows:

$$^{D}(V/K) = \left[^{D}K + (k_5/k_4)\left(^{D}k_3 + k_3/k_2\right)\right]/[1 + (k_5/k_4)(1 + k_3/k_2)] \tag{1}$$

where ^{D}K is the equilibrium isotope effect on the $ES \rightleftharpoons ES^*$ pre-decarboxylation equilibrium, $(k_5/k_4)(1+k_3/k_2)$ is the fractionation factor, and $^{D}k_3$ is the kinetic isotope effect on the proposed deuterium-sensitive individual step preceding decarboxylation. Using estimated values from model reactions and the calculated fractionation factor, the following equations are obtained:

$$1.00(\pm 0.06) = \left[0.87 + (k_5/k_4)\left(^{D}k_3 + k_3/k_2\right)\right]/1.84 \tag{2}$$

$$0.97 = (k_5/k_4)\left(^{D}k_3 + k_3/k_2\right) \tag{3}$$

When the value for the fractionation factor $0.84=(k_5/k_4)(1+k_3/k_2)$ is used with Eq. 3, one obtains:

$$1.15 = \left(^{D}k_3 + k_3/k_2\right)/(1 + k_3/k_2) \tag{4}$$

Since k_3/k_2 cannot be zero or negative, the only possible values for $^{D}k_3$ are greater than 1.15, which is contrary to the expectation in this mechanism (i.e., a secondary kinetic hydrogen isotope effect for sp^2 to sp^3 conversion should be less than 1). This analysis can only yield a reasonable value for $^{D}k_3$ if the values inserted into Eq. 1 are unreasonable. This analysis did not take into account the possibility that the proposed nucleophile added to C5 might dissociate concomitant with decarboxylation [27], in which case the proposed sp^3 to sp^2 re-hybridization would likely add another isotope-sensitive step to the analysis. However, the observed ODCase carbon isotope effect rises to 1.05 at low pH, making decarboxylation almost entirely rate-determining, and it would be difficult to envision how a pre-decarboxylation covalent step could occur with a carbon isotope effect approaching the expected intrinsic value. (By comparison, amino acid decarboxylases using covalent attachment by pyridoxal phosphate display observed carbon isotope effects distinctly lower than the intrinsic value.)

The pattern of carbon isotope effects with changing pH suggested the presence of a kinetically discernable proton-sensitive pre-decarboxylation step [24]. The observed carbon isotope effect did not rise to the level of the intrinsic isotope effect at the high pH extremes, but reached a maximum of about 1.035. This data is consistent with a mechanistic model in which ODCase protonates OMP, forming a kinetically observable intermediate, and the protonating enzyme functional group is titrated at high pH. This model, which includes the possible mechanisms involving proton transfer, was further supported by the later finding that D_2O as solvent for the *E. coli* ODCase reaction lowers the observed carbon isotope effect [28], a finding that was also made for the yeast enzyme [29].

Rishavy and Cleland [30] addressed the possibility of nitrogen isotope effects on the ODCase reaction, with attention to N1 and any possible mechanism involving a secondary nitrogen isotope effect, such as mechanism 1 in Fig. 1. A kinetic equation describing this mechanistic model would be as follows:

$$E + S \underset{k_2}{\overset{k_1}{\rightleftharpoons}} ES \underset{k_4}{\overset{k_3}{\rightleftharpoons}} ES^* \xrightarrow{k_5} E + P + CO_2 \quad (a)$$

where ES represents the Michaelis complex between ODCase and the substrate OMP, and ES^* represents the complex upon proton transfer from the enzyme to O2 of the substrate.

In this proposed mechanism, the protonation/deprotonation of O2 (k_3 and k_4) would be slightly kinetically sensitive to isotopic substitution at N1, as would the decarboxylation step (k_5). The equation relating the observed enzymatic ^{15}N isotope effect, $^{15}(V/K)$, with the isotope effects on the individual kinetic steps and pre-decarboxylation equilibrium is as follows:

$$15(V/K) = \left[\left(^{15}K\right)\left(^{15}k_5\right) + (k_5/k_4)\left(^{15}k_3 + k_3/k_2\right)\right]/[1 + (k_5/k_4)(1 + k_3/k_2)] \quad (5)$$

where ^{15}K is the ^{15}N equilibrium isotope effect on the ES to ES^* proton transfer equilibrium, and $^{15}k_5$ and $^{15}k_3$ are ^{15}N kinetic isotope effects on the individual steps in the reaction.

If ODCase in fact uses the O2 protonation mechanism, a solution of Eq. 5 must yield plausible values for the individual kinetic constants. Specifically, the values for $^{15}(V/K)$, ^{15}K, $^{15}k_5$ and $(k_5/k_4)(1+k_3/k_2)$, the commitment factor (c_f) for enzyme catalysis, obtained by direct measurement or comparison with model reactions, must yield a value of $^{15}k_3$ slightly less than 1, a predictable value for what would be an inverse secondary ^{15}N kinetic isotope effect on protonation at O2.

The authors [30] measured the nitrogen (N1) isotope effect $^{15}(V/K)$ for the ODCase reaction to be 1.0036 at pH 6.5, estimated c_f=0.885 at pH 6.5 from previous ^{13}C isotope effect studies, and used model reactions for values of ^{15}K and $^{15}k_5$. ^{15}K for protonation of pyridine (determined previously by Kurz et al. [31] to be 0.9793), was used to estimate ^{15}K for the ODCase reaction. A value of $^{15}k_5$=1.007, the value experimentally determined for decarboxylation of *N*-methylpicolinic acid, was used to estimate $^{15}k_5$ for the ODCase reaction. Using these values in Eq. 5 yields:

$$1.0036 = \left[(0.979)(1.007) + (k_5/k_4)\left(^{15}k_3 + k_3/k_2\right)\right]/[1 + 0.885] \quad (6)$$

$$.906 = (k_5/k_4)\left(^{15}k_3 + k_3/k_2\right) \quad (7)$$

From Eq. 7, and with $c_f=(k_5/k_4)(1+k_3/k_2)=0.885$, $^{15}k_3$ is necessarily greater than 1, an illogical result since protonation of O2 of OMP should be slightly

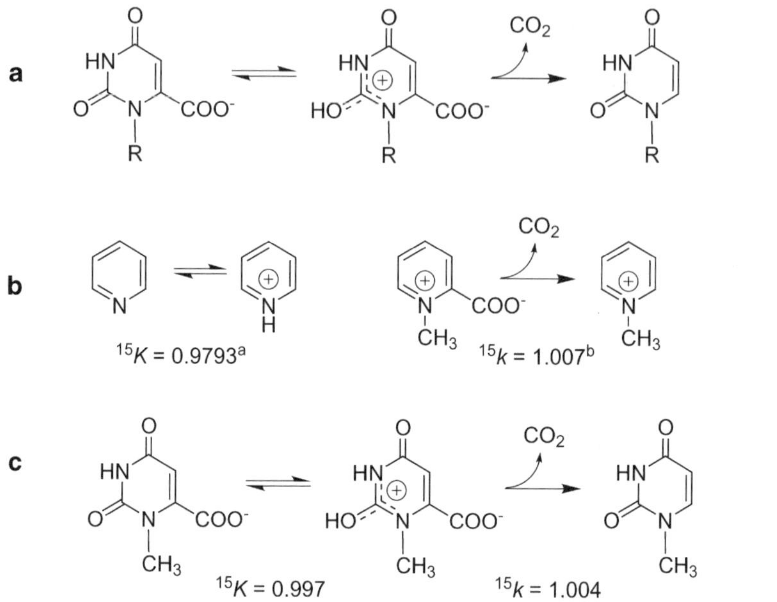

Fig. 3 **a** Proposed mechanism of ODCase-catalyzed decarboxylation of OMP by O2 protonation. Both the protonation and decarboxylation steps would be expected to be slightly sensitive to isotopic substitution at N1. **b** Model reactions used to assess the feasibility of the O2 protonation mechanism, or any mechanism with a pre-decarboxylation step that is isotopically sensitive at N1, and the measured N1 equilibrium and kinetic isotope effects. **a** Data from [31]. **b** Data from [30]. **c** Model reactions for which N1 equilibrium and kinetic isotope effects were determined using computational approaches, and the computed values [32]

faster with the heavier isotope ^{15}N present at N1. The authors reasoned that a mechanistic model with a pre-decarboxylation step insensitive to ^{15}N substitution more precisely fit the available data. In such a model, $^{15}K=1$, and solving Eq. 5 for $^{15}k_3$ with $^{15}K=1$ and the estimated value for $^{15}k_5$ yields $^{15}k_3=1.000$, as would be expected. This analysis was provided as evidence against any mechanism sensitive to N1 isotopic substitution, including the O2 protonation mechanism for ODCase.

Comparison of the model reactions used to those proposed for the ODCase reaction (Fig. 3) suggests that the model reactions are not ideal. With regard to ^{15}K, pyridine in solution is protonated directly at the nitrogen atom for which the isotope effect is being measured, but OMP at the enzyme active site (in the proposed mechanism 1 in Fig. 1) is not. ^{15}K for protonation of a heterocycle directly at N should be more substantial (i.e., lower, and farther from unity) than an analogous ^{15}K for protonation at an exocyclic site, since an important new, high-energy vibrational mode would be introduced by direct protonation of N. An assumed value of ^{15}K that is less than the actual value would lead to a solution of Eq. 5 for mechanism 1 proposed in Fig. 1 giving possible solutions for ${}^{15}k_3$ that are artificially high.

Phillips and Lee [32] used computational methods to obtain a value of 0.997 for the equilibrium nitrogen isotope effect at N1 for the O2 protonation of 1-methylorotate and a value of 1.004 for the kinetic nitrogen isotope effect at N1 for the decarboxylation of the proposed O2 protonated intermediate. Including these values in Eq. 5 yields:

$$1.0036 = \left[(0.997)(1.004) + (k_5/k_4)\left({}^{15}k_3 + k_3/k_2\right)\right]/[1 + 0.885] \quad (8)$$

$$.891 = (k_5/k_4)\left({}^{15}k_3 + k_3/k_2\right) \quad (9)$$

With $c_f=(k_5/k_4)(1+k_3/k_2)=0.885$, a solution for ${}^{15}k_3<1$ is possible if the values for ${}^{15}(V/K)$, ${}^{15}K$, ${}^{15}k_5$ and c_f are extended to reasonable limits of error, and is dependent on the exact values inserted into Eq. 5 to the third decimal place for confirmation or refutation. An attempted confirmation of ${}^{15}k_3=1.000$ for a mechanism involving an N1-independent pre-decarboxylation step is also dependent on exact kinetic values. Using nitrogen isotope effects to discriminate between an O2-protonation pre-decarboxylation step with ${}^{15}K=0.997$ and an N1-independent pre-decarboxylation step with ${}^{15}K=1.000$ would seem to be a formidable task.

5 Alternate Substrates for ODCase

Alteration of enzyme substrates and observation of the enzyme's activity toward the altered substrate has been a widespread enzymological approach. Shostak and Jones [33] synthesized a series of OMP analogs, primarily designed to address the possibility of the O2-protonation mechanism. Thio substitution for a carbonyl oxygen often renders the altered substrate incapable, or poorly capable, of undergoing enzyme catalysis when the enzyme makes a critical contact with the carbonyl; for example, acetyldithio-coenzyme A is a poor substrate for citrate synthase [34]. With this pretext, Shostak and Jones found 4-thioOMP to be an ODCase substrate with kinetic values relatively unchanged from those for OMP. The UV absorbance change in this decarboxylation, using a wavelength higher than that for the absorbance of total protein, provided a novel spectroscopic assay for ODCase [35]. By contrast, this group's preparation of 2-thioOMP appeared not to be decarboxylated by ODCase, and showed a modest degree of inhibition.

The substrate utilization of 2-thioOMP was re-examined with the intent of determining whether a small extent of decarboxylation by ODCase might be detectable with an assay more sensitive than the spectroscopic assay previously utilized [36]. An amount of 2-thioUMP that would have been present if 2-thioOMP were decarboxylated at a rate 10^{-7} times that of the normal substrate was undetectable. Additionally, the new preparation of 50 μM 2-thioOMP did not inhibit the ODCase reaction toward OMP, revealing that thio substitution at the 2-carbonyl position of OMP essentially eliminates binding of the altered substrate. This finding was especially curious in light of the observation that 2-thio substitution of UMP does not appreciably alter the dissociation constant of the product [18].

5-FluoroOMP was found to be decarboxylated by ODCase with kinetic constants comparable to those for OMP [33]; in fact, the k_{cat} for 5-fluoroOMP was measured to be 30 times greater than that for OMP. If ODCase uses an electrophilic substitution mechanism, protonating the substrate somewhere along the 5,6 double bond (Fig. 1, mechanism 2 or 3), an electrophilic substitution pre-decarboxylation step would almost certainly be substantially slower with 5-fluoroOMP. In order to compensate for this slower pre-decarboxylation step and yield an overall catalytic mechanism with an increased k_{cat}, the decarboxylation step would apparently have to be several orders of magnitude faster with 5-fluoroOMP.

The contributions of the substrate phosphoryl group and phosphoribosyl group were measured for the human and yeast enzymes [37, 38]. ODCase demonstrates measurable but highly reduced activity toward orotidine and orotate, indicating substantial contributions to catalysis made by the phosphoryl group and phosphoribosyl group, respectively. The contribution of the phosphoryl group in accelerating the progress from E·S to E·TS by the human enzyme can be assessed by comparing the k_{cat} values for OMP and orotidine; k_{cat} for OMP is greater than that for orotidine by a factor of 10^5. With the assumption that k_{uncat} is the same for OMP and orotidine, this comparison can be expressed as a comparison of ratios of K_m/K_{TS}. (Comparison of K_{TS} alone for OMP and orotidine would be a comparison of the affinity of free ODCase for the transition states of decarboxylation for these two substrates, a slightly different comparison.)

Removal of the phosphoryl group from an enzyme's substrate and observation of diminished kinetics has been seen with other enzymes, notably with triose phosphate isomerase (TIM) [39]. The k_{cat} for TIM using glyceraldehyde as substrate is diminished by a factor of approximately 10^9; the rate of catalysis for the dephosphorylated substrate is comparable to the rate of the uncatalyzed reaction. The phosphoryl groups of TIM ligands contribute to the structure of the active site: the phosphoryl group of the inhibitor phosphoglycolohydroxamate is bridged to a water molecule and subsequently to Lys12 of the yeast enzyme, which in turn contacts O2 of the triose phosphate during catalysis [40]. The decrease in k_{cat} for ODCase from phosphorylated to dephosphorylated substrates is not as large as that for TIM, but is still suggestive of a direct role for the phosphoryl group in catalysis.

6 Reconciling Enzymological Data with Structural Data: Is ODCase a Two-faced Enzyme?

ODCase experienced its most tumultuous period of popularity in 2000, with the nearly simultaneous publication of four crystal structures [9, 10, 11, 41] and a five-page article in C&E News [42]. The structures of ODCases from four microbial sources are quite similar and suggest identical mechanisms. The amino group of Lys93 of the yeast enzyme [41] (corresponding to Lys62 of the *B. subtilis* enzyme [9], Lys72 of the *M. thermoautotrophicum* enzyme [10] and Lys73 of the *E. coli* enzyme [11]) is seen in the crystal structures to be in close proximity with O6 of the ligand BMP (Fig. 4a). (In other structures with UMP or 6-azaUMP, the amino group of the active site lysine is in close proximity with C6 or N6.) By superimposing the structure of OMP onto the structures of the ligands, a binding orientation for OMP with the substrate carboxylate proximal to the catalytic lysine was inferred. New mechanisms for ODCase activity, most notably mechanism 2 (Fig. 1), were proposed and included the possibility of substrate destabilization by Asp91 and Asp96 (the latter from the second monomer in the dimer), or their corresponding positions in the bacterial enzymes. Most of the recent computational studies on the possible mechanisms of ODCase [10, 43, 44, 45] have been based on this binding orientation. The structures of ODCase in complex with inhibitors would seem to eliminate the mechanism involving Lys93 protonation of O2, since the pyrimidine rings are turned with the O4′-C1′-N1-C2 dihedral angle in a configuration opposite from that needed to bring together O2 and Lys93.

The symmetry of the pyrimidine ring of BMP, and its consequences for binding to ODCase, had been noted previously [17, 19]. Regardless of the binding orientation of OMP, if an active site cation were present within ODCase, contacting any side of the OMP pyrimidine ring during catalysis, the anionic C4-C6 side of BMP would be expected to have a more favorable contact with that cationic functional group than would the C2-C4 side. Similarly, UMP and 6-azaUMP, without the steric consideration of the carboxylate of OMP, could presumably fit into an active site for OMP either with the ring atoms superimposed, or with the glycosidic bond rotated from that structure by about 180°.

If the structure of ODCase complexed with BMP in Fig. 4a is modified by substituting a carboxyl group for the ionized hydroxyl at C6, BMP would be replaced with OMP (Fig. 4b). In this structure, there are no close active site contacts with O2; Gln215, highlighted at left, is the only close contact, yet its removal by mutagenesis does not affect ODCase activity [46]. This possible active site structure for OMP binding, which has not been optimized, does not provide an apparent explanation for the inactivity of ODCase toward 2-thioOMP. Substitution of the carbonyl O2 in UMP with S predictably does

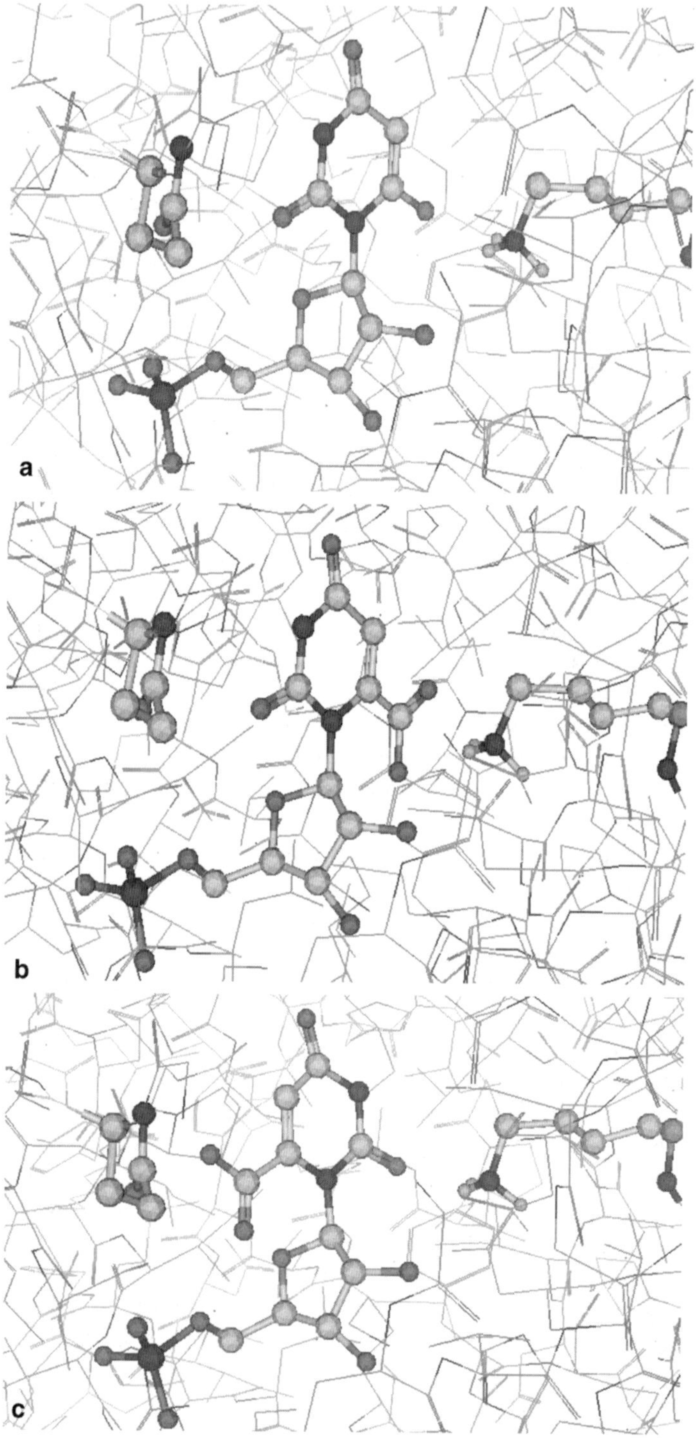
a
b
c

not alter the affinity of ODCase for the altered ligand, since there is no close contact of O2 or S2, but substitution of the carbonyl O2 in OMP with S essentially eliminates binding. The Q215A ODCase mutant is also inactive toward 2-thioOMP (B. G. Miller and R. Wolfenden, personal communication), so the possibility that Gln215 obstructs 2-thioOMP binding does not seem satisfactory.

If the above modification of Fig. 4a includes the further modification of rotating the O4′-C1′-N1-C2 dihedral angle by 180°, a binding model (Fig. 4c) emerges that seems to account for much of the enzymological data. Lys93 would be within protonation distance of O2, which could lead to the formation of a kinetically discernable protonated intermediate, as suggested by the isotope effect data. Substitution of sulfur for oxygen at O2 could result in unacceptable contacts between the enzyme and the alternate substrate, in agreement with the enzymological data. Lys93 and Asp91 are within 3 Å of O2, and substitution of O2 with sulfur might not allow the altered substrate to bind due to these close contacts. The carboxylate group, removed during catalysis, is within proximity of the substrate phosphoryl group. The oxygen atoms of the carboxylate could be within 3–5 Å of the phosphoryl oxygens; this is similar to the charge repulsion contacts inferred between the substrate carboxylate and an enzyme glutamyl side chain in the pyruvoyl-dependent histidine decarboxylase [47]. The role of the phosphoryl group in catalysis could be to assist in promoting electron flow off the carboxylate to liberate CO_2, in accord with the diminished activity of ODCase toward orotidine. The necessary protonation states of Lys93 and the substrate phosphoryl group inferred from this model—Lys93 must be protonated below its pK_a and the phosphoryl group would likely be dianionic—are consistent with kinetic measurements of the pH dependence [48]. The source of the apparently lowered pK_a for Lys93 from a usual value of about 10 and its pro-

◄

Fig. 4a–c Images of the yeast ODCase crystal structure in complex with BMP, imported from the Protein Data Bank file 1DQX using the Molecular Operating Environment (MOE)1, version 2002.03, Chemical Computing Group, Montreal, QC. BMP and the side chains of residues Gln215 (*left*) and Lys93 (*right*) have been highlighted. Leu213 and Ile97 (B subunit) have been deleted for clarity. **a** Unaltered structure of ODCase-BMP, showing proximity of O6 to the amino group of Lys93. **b** Structure of ODCase-BMP altered by changing O6 of BMP to a carboxylate; the dihedral angle of the C6-carboxyl bond was arbitrarily set at 0° using MOE. Alteration of the ligand results in the hypothetical construction of OMP at the active site with C6 and the carboxylate oriented toward the amino group of Lys93. **c** Structure of ODCase-BMP altered by changing O6 of BMP to a carboxylate and rotating the glycosidic bond by 180°; the dihedral angle of the C6-carboxyl bond was arbitrarily set at 0° using MOE. Alteration of the ligand results in the hypothetical construction of OMP at the active site with O2 oriented toward the amino group of Lys93

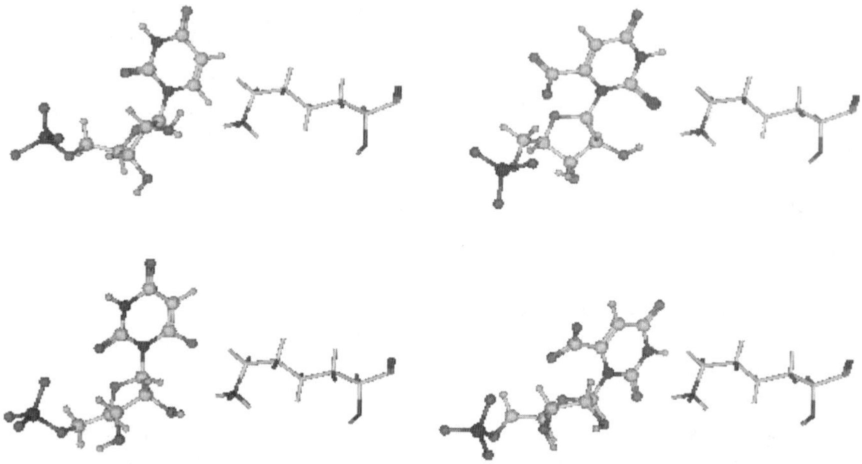

Fig. 5 Results of molecular docking simulations of UMP (*upper left*), BMP (*lower left*), OMP (*lower right*), and 2-thioOMP (*upper right*) at the active site of yeast ODCase, with reference to Lys93, using MOE

posed ability to protonate the weakly basic O2 of the substrate still need explaining.

Preliminary data from molecular modeling studies (A. R. Lawrence and J. A. Smiley, in preparation) indicate that OMP binds to ODCase with O2 in close proximity with Lys93 of the yeast enzyme (Fig. 5), with the pyrimidine ring rotated at the glycosidic bond essentially 180° from that seen with UMP and BMP. In this analysis, all calculations were performed using the Molecular Operating Environment (MOE) 1, version 2002.03, available from Chemical Computing Group, Montreal, QC. The crystal structure 1DQX for yeast ODCase [41] complexed to BMP was imported into MOE and the associated BMP ligands were removed. Using MOE-Dock, a docking box of dimension 44×44×44 Å with 0.375 Å grid spacing was centered at the active site in 1DQX.A. Amino acid residues from 1DQX.B that contribute to the "A" active site were included in this docking box. Free UMP, BMP, OMP, and 2-thioOMP were docked into the fixed receptor using a Tabu Search global search algorithm (which prevents re-visitation of previously encountered forbidden structures). In each experiment, 25 runs were performed with 1,000 steps per run. The docking results were examined and the structure with the lowest calculated total energy of interaction was brought into the receptor.

Figure 5 shows modeling results consistent with the mechanism involving Lys93 protonation of O2, and consistent with enzymological data. UMP and BMP are projected to approach Lys93 of the active site with a binding orientation consistent with that seen in the crystal structures. By contrast, OMP is predicted by this analysis to bind with the glycosidic bond rotated by roughly 180°, and with O2 within 3.4 Å of the Lys93 amino group. The projected binding of 2-thioOMP is contorted, especially with respect to the ribosyl and phosphoryl groups, suggesting that this substrate analog does not bind well to the ODCase active site, in accord with the lack of activity of ODCase toward 2-thioOMP and the lack of 2-thioOMP inhibition of ODCase activity toward the natural substrate.

Alternate binding orientations for nearly or even vaguely symmetrical heterocycles at enzyme active sites is not uncommon in enzymology. An intriguing example is found in the active site of UDP-galactose-4-epimerase, which catalyzes the isomerization of UDP-galactose and UDP-glucose [49]. Not only is it proposed that the hexose moiety rotates during transient oxidation and reduction, but also that the enzyme bound to its non-dissociating nicotinamide co-factor orients the NAD^+ form in the *syn* conformation and the NADH form in the *anti* conformation. The model for ODCase catalysis involving Lys93 protonation of O2 of OMP does not include rotating the pyrimidine ring during or after catalysis, but simply proposes that UMP is generated by decarboxylation of OMP with O2 facing Lys93, dissociates from this conformation, and binds in a reverse pyrimidine orientation when introduced back into the active site as an inhibitor.

Support for this model or others is currently being sought with a combination of spectroscopic and kinetic studies using various inhibitors and substrates.

Acknowledgements Research in the author's laboratory is funded by NIH Grant R15 GM63504-01, the Youngstown State University Presidential Academic Center for Excellence in Research, and the Ohio Board of Regents Hayes Investment Fund. The ongoing collaboration with Raelene Lawrence, Chemical Computing Group , Montreal, is gratefully acknowledged, as are collaborations with other ODCase investigators.

References

1. O'Leary MH (1992) The Enzymes 20:235
2. Abell LM, O'Leary MH (1988) Biochemistry 27:5927
3. O'Leary MH (1976) Biochem Biophys Res Comm 73:614
4. Waldrop GL, Braxton BF, Urbauer JL, Cleland WW, Kiick DM (1994) Biochemistry 33:5262
5. Miller BG, Traut TW, Wolfenden R (1998) J Am Chem Soc 120:2666
6. Miller BG, Smiley JA, Short SA, Wolfenden R (1999) J Biol Chem 274:23841
7. Cui W, DeWitt JG, Miller SM, Wu W (1999) Biochem Biophys Res Comm 259:133
8. Beak P, Siegel B (1976) J Am Chem Soc 98:3601
9. Appleby TC, Kinsland C, Begley TP, Ealick SE (2000) Proc Natl Acad Sci USA 97:2005
10. Wu N, Mo Y, Gao J, Pai EF (2000) Proc Natl Acad Sci USA 97:2017
11. Harris P, Navarro-Poulsen JC, Jensen KF, Larsen S (2000) Biochemistry 39:4217
12. Lee TS, Chung LT, Chodera JD, Kollman PA (2001) J Am Chem Soc 123:12837

13. Silverman RB, Groziak MP (1982) J Am Chem Soc 104:6434
14. Lee JK, Houk KN (1997) Science 276:942
15. Radzicka A, Wolfenden R (1995) Science 267:90
16. Brody RS, Westheimer FH (1979) J Biol Chem 254:4238
17. Levine HL, Brody RS, Westheimer FH (1980) Biochemistry 19:4993
18. Smiley JA, Saleh L (1999) Bioorg Chem 27:297
19. Smiley JA, Jones ME (1992) Biochemistry 31:12162
20. Radford A, Dix NIM (1998) Genome 30:501
21. Kimsey HH, Kaiser D (1992) J Biol Chem 267:819
22. Acheson SA, Bell JB, Jones ME, Wolfenden R (1990) Biochemistry 29:3198
23. Kirsch JF (1977) In: Cleland WW, O'Leary MH, Northrop DB (eds) Isotope effects on enzyme-catalyzed reactions. University Park Press, Baltimore, p 100
24. Smiley JA, Paneth P, O'Leary MH, Bell JB, Jones ME (1991) Biochemistry 30:6216
25. Northrop DB (1977) In: Cleland WW, O'Leary MH, Northrop DB (eds) Isotope effects on enzyme-catalyzed reactions. University Park Press, Baltimore, p 122
26. Cook PF, Cleland WW (1981) Biochemistry 20:1790
27. O'Leary MH (1989) Ann Rev Biochem 58:377
28. Ehrlich JI, Hwang CC, Cook PF, Blanchard JS (1999) J Am Chem Soc 121:6966
29. Smiley JA (1991) PhD thesis, University of North Carolina
30. Rishavy MA, Cleland WW (2000) Biochemistry 39:4569
31. Kurz JL, Daniels MW, Cook KS, Nasr MM (1986) J Phys Chem 90:5357
32. Phillips LM, Lee JK (2001) J Am Chem Soc 123:12067
33. Shostak K, Jones ME (1992) Biochemistry 31:12155
34. Wlassics ID, Anderson VE (1989) Biochemistry 28:1627
35. Shostak K, Christopherson RI, Jones ME (1990) Anal Biochem 191:365
36. Smiley JA, Hay KM, Levison BS (2001) Bioorg Chem 29:96
37. Miller BG, Traut TW, Wolfenden R (1998) Bioorg Chem 26:283
38. Miller BG, Butterfoss GL, Short SA, Wolfenden R (2001) Biochemistry 40:6227
39. Amyes TL, O'Donoghue AC, Richard JP (2001) J Am Chem Soc 123:11325
40. Davenport RC, Bash PA, Seaton BA, Karplus M, Petsko GA, Ringe D (1991) Biochemistry 30:5821
41. Miller BG, Hassell AM, Wolfenden R, Milburn MV, Short SA (2000) Proc Natl Acad Sci USA 97:2011
42. Rouhi AM (2000) Chem Eng News 78:42
43. Warshel A, Strajbl M, Villa J, Florian J (2000) Biochemistry 39:14728
44. Lundberg M, Blomberg MRA, Siegbahn PEM (2002) J Mol Model 8:119
45. Hur S, Bruice TC (2002) Proc Natl Acad Sic USA 99:9668
46. Miller BG, Snider MJ, Wolfenden R, Short SA (2001) J Biol Chem 276:15174
47. Gallagher T, Snell EE, Hackert ML (1989) J Biol Chem 264:12737
48. Porter DJT, Short SA (2000) Biochemistry 39:11788
49. Thoden JB, Frey PA, Holden HM (1996) Biochemistry 35:2557

Top Curr Chem (2004) 238:79–112
DOI 10.1007/b94540

Developing Active Site Models of ODCase— from Large Quantum Models to a QM/MM Approach

Marcus Lundberg (✉) · Margareta R. A. Blomberg · Per E. M. Siegbahn

Department of Physics, Stockholm Center for Physics, Astronomy and Biotechnology, Stockholm University, 106 91 Stockholm, Sweden
marc@physto.se

Abstract The catalytic mechanism of orotidine monophosphate decarboxylase (ODCase) has been modeled using density functional theory with the B3LYP functional. Barriers for three different mechanisms have been calculated using large QM and QM/MM models. A concerted protonation mechanism where TS stabilization is provided only by the positive Lys93 has a high barrier around 35 kcal/mol. QM/MM calculations confirm the results obtained using QM models. For a base protonation mechanism, O2 protonation gives a barrier for decarboxylation of 26 kcal/mol. Extensions to this QM model indicate that the cost of protonation may be underestimated and the support for the base protonation mechanism is uncertain. An initial QM/MM investigation of a stepwise mechanism, where water molecules seem to play an important role for TS stabilization, gives the most promising results with an estimated barrier of 22 kcal/mol.

Keywords Enzymatic mechanism · Active site · ODCase · Density functional theory · QM/MM

Abbreviations

DFT	Density functional theory
MM	Molecular mechanics
OMP	Orotidine 5′-monophosphate
ODCase	Orotidine 5′-monophosphate decarboxylase
QM	Quantum mechanics
QM/MM	Quantum mechanics/molecular mechanics
TS	Transition state

1 Introduction

Enzymes catalyze many reactions far better than man-made catalysts. The goal of theoretical studies of enzyme mechanisms is to understand how they achieve their amazing catalytic power. In this respect, orotidine 5′-monophosphate decarboxylase (ODCase) is one of the most fascinating enzymes. It increases the rate of decarboxylation of its substrate orotidine 5′-monophosphate (OMP) (see Fig. 1) by 17 orders of magnitude. This rate acceleration is unmatched by other known enzymes [1]. Analyzing the mechanism of ODCase can therefore be an important step in understanding enzyme catalysis on a more general level.

Mechanisms of enzyme catalysis can be attacked by theoretical methods. However, the complexity of enzymes puts severe requirements on the theoretical modeling. There are practical limits as to the size of the models and it can be quite difficult to design the models to be used. A successful model

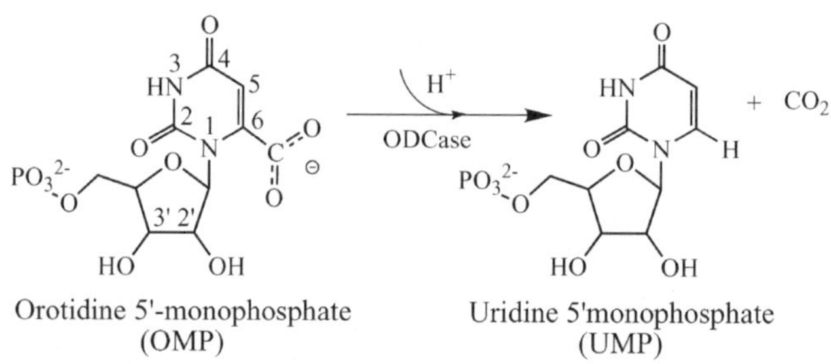

Fig. 1 General scheme of the substrate reaction in ODCase

has to include all effects relevant for the mechanism of the enzyme with sufficient accuracy. At present, some computational methods give the required accuracy but can only treat a very small part of the enzyme. Other methods can treat the entire enzyme, but these methods are in general not accurate enough. Despite these difficulties, different theoretical approaches have been applied to the study of enzyme mechanisms with considerable success. The present review will describe one specific computational approach, an accurate quantum mechanical modeling of the enzyme active site, and will discuss how this approach can be used to learn more about the catalytic power of the ODCase enzyme.

1.1 Mechanisms Proposed for the Enzymatic Decarboxylation in ODCase

As can be seen in Fig. 1 one main event of the substrate reaction in ODCase is that the carboxylate group attached to the pyrimidine ring leaves to form carbon dioxide. During this reaction, the negative charge of the carboxylate is transferred to the pyrimidine ring. In other known decarboxylations, this negative charge is stabilized by delocalization either into a π-system or by a cofactor present at the active site, but this is not the case in ODCase [2, 3, 4]. The main task of a computational study of ODCase is therefore to understand by which means this enzyme accomplishes the stabilization of the negative charge.

The other main event of the substrate reaction is that a proton replaces the carboxylate on C6 of the pyrimidine ring, see Fig. 1. This proton can approach the substrate either before, during or after the C–C bond is cleaved, and it will contribute to the charge stabilization to different degrees in the different cases. The three alternatives for the protonation step give rise to three different classes of reaction mechanisms for the ODCase enzyme, which will be discussed below in a separate section for each type of mechanism. The first mechanism discussed is the one where the proton attacks C6 at the same time as the C–C bond cleaves. This mechanism is referred to as the concerted mechanism, and in this case it is the incoming proton that stabilizes the negative charge on the pyrimidine ring. The second alternative is that a proton is already attached to the pyrimidine ring of the substrate before the C–C bond starts to cleave. Several different alternatives for the site of protonation on the pyrimidine ring have been proposed, the O2, O4 and C5 positions (for atomic labels see Fig. 1), [5, 6, 7]. These proposals will be referred to as the base protonation mechanisms. The third alternative is that a proton comes in only after the C–C bond is more or less cleaved. In this case, the stabilization of the negative charge on the pyrimidine ring during the reaction has to use other means than a single proton. The three mechanisms are schematically shown in Fig. 2 and this review will discuss all three classes of mechanisms, although the focus will be on the first two.

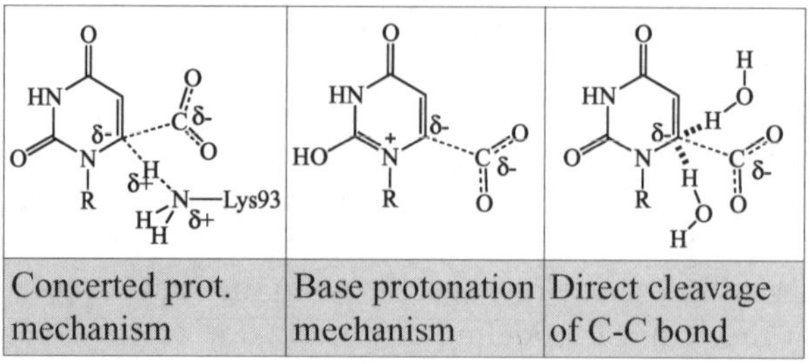

Fig. 2 Schematic illustration of the three different mechanisms treated in this review. The base protonation mechanism treats protonation of O2 as well as all other sites of the pyrimidine ring

1.2 Modeling Enzymatic Reactions Using Accurate QM Models

How can a theoretical method decide between proposed mechanisms, and how can the origin of the enzymatic power be identified? This review will try to answer these questions for one particular theoretical approach, the one where an active site model is treated by accurate quantum mechanical (QM) methods. The main idea in the QM active site approach is to make sure that the computational results have the required accuracy. During the last decade the accuracy of density functional methods (DFT) has been dramatically improved, and in particular the hybrid B3LYP functional has achieved a remarkable accuracy [8, 9]. The use of DFT has also made it possible to treat dramatically larger molecular systems than can be done with conventional wave-function methods of similar accuracy. In spite of this important development, DFT models have usually been limited to 50–60 atoms, but more recently systems with more than 100 atoms have been treated efficiently. Still, even 100 atoms is a very small part of the total number of 8,300 atoms in yeast ODCase, not counting hydrogens or surrounding water molecules. Thus a very severe selection has to be made when the enzyme model is set up, and an important task is to select the residues required to solve the mechanism and to analyze all important contributions.

One assumption in the QM modeling is that the active site is the only important part for the substrate reaction. From experimental data it is known that most invariant amino acids are located very close to the active site of the enzyme. Out of the eight invariant amino acids in ODCase [10], six make direct contact with the substrate and Asp91 (yeast enzyme numbering is used throughout this contribution) binds to two other invariant amino acids [11, 12, 13, 14]. Asp91 is part of a network of four invariant and charged amino acids: Lys59, Asp91, Lys93, and Asp96B. This network is located close to the leaving carboxylate (see Fig. 3) and mutation experiments show that they are important for enzyme activity [15]. Mutation studies have also

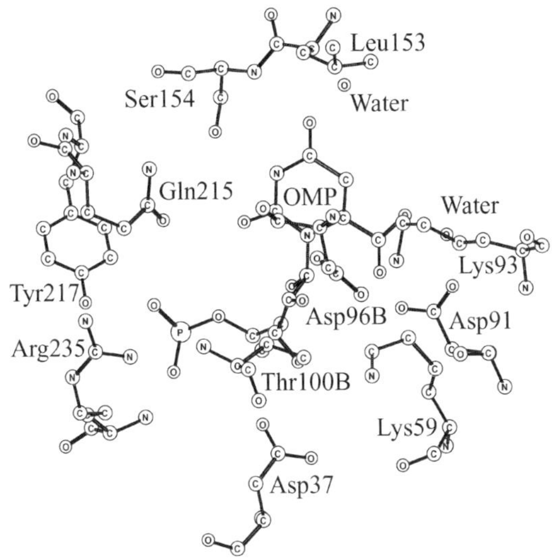

Fig. 3 Picture of the active site in ODCase showing only selected amino acids. The geometry of the active site is taken from the X-ray structure 1DQX and the carboxylate group has been added to the inhibitor BMP. Selected residues have either been shown to be important for catalysis or play important roles in the computational models of this review. No hydrogens are included in the figure

shown effects of highly conserved amino acids in the active site, e.g., Thr100 [16]. Unfortunately, including all invariant and highly conserved amino acids in a single model is not manageable. For ODCase, the model would consist of more than 200 atoms. A selection is required.

Such a selection cannot be accomplished without a hypothesis for how the reaction proceeds and which amino acids are most important for the mechanism. This hypothesis should be based on known experimental data, and should include structures for the reactant, the product, possible intermediates, and transition states. Amino acids from the enzyme should be included if they are expected to have any type of differential effects on the different structures. Every hypothesis is tested by a comparison between computational and experimental results. The most important computational result is the energy diagram for the stationary points on the potential energy surface. This energy surface is compared to key experimental data, e.g., turnover rates and observed intermediates. In the case of ODCase k_{cat} is 20 s^{-1} [17], which by transition state theory corresponds to a reaction barrier of 15 kcal/mol. This means that the highest barrier on the potential energy surface should be about 15 kcal/mol, for a mechanism to be valid. There is no experimental data regarding intermediates. In solution, the measured reaction barrier for decarboxylation of the same substrate is 38 kcal/mol. Thus, the enzyme manages to lower the barrier by an amazing 23 kcal/mol.

If a QM model of the active site reproduces the barrier and other experimental observations, it is likely that the reaction actually proceeds via this particular mechanism. An important aspect of a theoretical modeling of enzyme mechanisms is that a model that turns out to give a successful description of the reaction can later be decomposed to identify which amino acids make the critical contributions to the catalytic power of the enzyme. The appeal of the limited active site model is that if the effect of the enzyme is reproduced, it can be argued that no other important effects are required. As the accuracy is believed to be sufficient, a correct barrier does not leave any room for any other effects. If the model fails to reproduce the enzymatic effect there are two alternative interpretations: either the suggested mechanism is actually not feasible, or the model does not include all important effects, and therefore has to be extended or modified.

The present methodology, using accurate QM models that treat only a rather small part of the active site, has met considerable success during the last few years. The authors of the present review have used the method mainly for metalloenzymes [18, 19, 20] but have also applied the methodology to ODCase [21]. That study treated the concerted reaction mechanism and the base protonation mechanism with O2 as protonation site. The present review includes those results but also presents significant extensions to the modeling of these mechanisms. In addition, results from investigations of other base protonation mechanisms, and the mechanism where the C–C bond is cleaved prior to protonation are also presented.

1.3
Adding the Surrounding Protein to an Accurate QM Model

The parts of the enzyme that are excluded from the QM model can still be treated in this approach, although in a more approximate way. A common and simple method is to use a homogenous dielectric medium, which intends to mimic the polarizability of the surroundings. Also, steric effects of the surrounding protein can be treated in a simple way by keeping some atoms frozen in their positions in the X-ray structure. A more advanced way to include polarizability and steric effects of the surrounding protein is to actually include the atoms excluded in the QM model and treat them at a lower level of theory, e.g., molecular mechanics (MM). The total treatment is then a combined quantum mechanics/molecular mechanics (QM/MM) method. This treatment can mimic the inhomogeneous properties of the enzyme and describe possible long-range electrostatic effects. In principle, it can also give better estimates of possible steric effects on the reaction barrier. What remains to be seen is whether these effects are important in the case of ODCase. The drawback of QM/MM methods is the increased computational effort and the risk of introducing artificial effects from the MM part. Due to the size of the MM part, these effects can be difficult to monitor and avoid.

This contribution reviews computational results for three classes of reaction mechanisms proposed for ODCase. Firstly, the mechanism that assumes protonation of C6 concerted with decarboxylation is described. Secondly, the base protonation mechanisms are reviewed. Finally, a shorter treatment is given of a reaction mechanism where the C–C bond is broken before the proton attaches to the base. All values in the review are obtained by the use of QM models of the active site. Effects of different residues on the reaction barrier are analyzed when going from small to large QM models. A QM/MM treatment is applied to each mechanism to see whether this treatment has any major effect on the calculated results. The goal of the review is to provide information regarding the activity of ODCase and to shed light on the requirements on QM models that are applied to enzymatic systems.

2 Computational Details

In the introduction, it was stated that accuracy is a key factor in the use of QM active site models. In this review, all calculations are made using the DFT hybrid functional B3LYP [8, 9]. The accuracy of the B3LYP method has been estimated using a benchmark test, the extended G3 set (376 entries). Including only the 301 entries that are most relevant for enzymatic reactions, where only a few bonds are formed or broken, the B3LYP functional has an average error of 3.29 kcal/mol [22]. To further estimate the accuracy of the calculated B3LYP barriers in the ODCase mechanisms, accurate G2MS calculations [23] were performed on two small models. Compared to the G2MS calculations, B3LYP gives a 4 kcal/mol lower barrier for a concerted TS (concerted protonation mechanism) and a 1 kcal/mol lower barrier for cleavage of the C6–carboxylate bond without proton transfer. The errors are in the range of a few kcal/mol and the B3LYP treatment is therefore acceptable for the kind of problems investigated in this review.

After the model has been selected for each problem, the coordinates of the residues are extracted from the PDB-structure 1DQX (yeast enzyme) [12]. Since the X-ray structure is only available with an inhibitor instead of the real substrate, the carboxylate group is simply added to the inhibitor BMP (6-hydroxyuridine 5′phosphate) in the X-ray structure. The long amino acid side chains are usually removed from the model since they do not have any important effects on the calculated barriers [24]. Starting from these structures, the calculations are performed in several steps.

Geometries are first optimized using a double zeta basis set, usually the d95 basis set [25]. Optimizing with a small basis set is sufficient since the final energy is rather insensitive to the quality of the geometry optimization [26]. To check the influence of the neglect of polarization functions in the optimization, a reaction barrier was calculated following optimization with the 6–31G(d,p) basis set instead of the d95 basis set. Only a minor change in the barrier height (−0.2 kcal/mol) was found. The d95 basis set is used also for the Hessian calculations, i.e., second derivatives of the energy with respect to the nuclear coordinates. Optimized equilibrium structures are accepted if the Hessian only has positive eigenvalues. Transition states (TS) are also obtained by full optimizations and are characterized by a single negative eigenvalue in the Hessian. The Hessians are also used to estimate zero-point, thermal, and entropy effects on the relative energies. Following a geometry optimization, the electronic energy is calculated using a larger basis set with polarization functions added to all atoms and diffuse functions added to the heavy atoms (d95+(2d,2p), d95+(d,p), and 6–311+G(d,p) have all been used). QM calculations in this review are performed with the programs Jaguar [27] or Gaussian98 [28].

The part of the protein that is not explicitly included in the model is treated as a homogenous medium with a dielectric constant of 4.0. Corrections for solvent effects are calculated using either SCIPCM [29, 30] or CPCM polarizable conductor model (Cosmo) [31, 32] or the Poisson–Boltzman solver in Jaguar [33]. In some models, the position of one atom in each residue is frozen to keep the residues close to the positions they have in the X-ray structure of the protein. Whenever any atom is kept frozen during the geometry optimization, this is indicated in the figure illustrating the model. In models with frozen atoms it is generally not possible to get accurate thermal corrections from the Hessians, and the energies reported for these models either do not include thermal corrections at all or use relevant values from smaller models without frozen coordinates.

An alternative way to treat the surrounding protein is by including it in a MM description. All QM/MM calculations in this review are made with the program Qsite [34, 35, 36]. In this program the QM/MM interface is built by frozen localized molecular orbitals. Only predefined cuts between MM and QM parts are allowed. The QM parts of the QM/MM models therefore include the entire residue side chains. The MM part in a Qsite approach is treated with the OPLS–AA force field [37]. A dimer of the yeast enzyme consists of more than 8,000 atoms, which is the maximum number of atoms that Qsite can handle. To reduce the size of the MM part below this limit, parts of the second chain (B) that are more than 10 Å away from the substrate are removed as well as the entire C and D chains. To keep the overall structure of the enzyme, approximately 40 atoms located close to the surface are kept frozen during the optimizations. The QM/MM treatment does not include dynamics and minima are located by optimizations starting from the X-ray structure.

Solvation calculations for the QM/MM treatment did not converge. The comparison between QM and QM/MM barriers therefore neglect solvent effects for both treatments. This means that adding MM is not considered as solvation. Contributions to the dielectric constant of the protein come from surrounding water molecules, electronic polarization of the residues, and possible movements of the residues. Of these, electronic polarization outside the QM part is not accounted for, and the effect of the surrounding water molecules is totally neglected in the present treatment. The only polarization effect that remains is the movement of the MM part but these effects are quite small when similar MM minima are used for the two structures.

3 Concerted Protonation Mechanism

3.1 QM Model with Lys93 as Proton Donator

From the X-ray structure of ODCase it is expected that a lysine residue (Lys93) should be particularly important for catalysis since it interacts directly with the carboxylate. One possibility for decarboxylation is a single step reaction where C6 of the pyrimidine ring is protonated by Lys93 concerted with cleavage of the bond between C6 and the carboxylate. In this mechanism, the negative charge developing on the ring is directly stabilized by the positive lysine. The first group to propose this possibility was Appleby et al. [11].

In the introduction it was stressed that two keys to successful QM modeling is model selection and formulation of reaction hypothesis. In the concerted reaction mechanism, the hypothesis is simple and straightforward. The smallest model that can treat the concerted reaction consists of just Lys93 and the relevant parts of the substrate (Model 1 in Fig. 4). In Model 1, both the lysine and the substrate are neutral. There is no important difference between this model and a model with a protonated lysine and a depro-

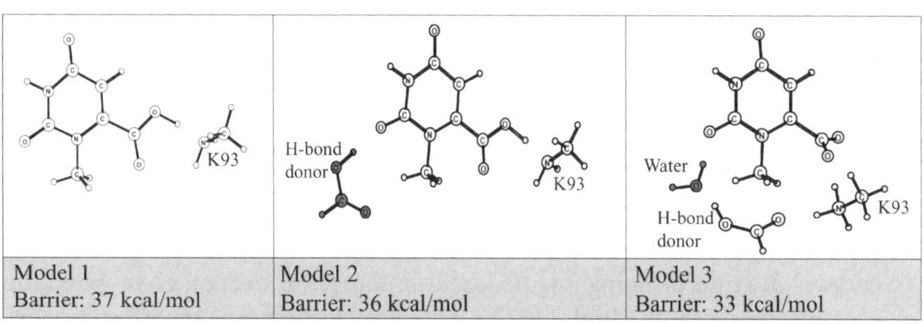

Fig. 4 QM models used for investigations of the concerted reaction mechanism where Lys93 protonates C6. This series of models shows the effects of hydrogen bonds to O2. Extensions made to a previous model in the series are *shaded* to facilitate comparison between models. Note that the structures are not always consistent with the X-ray structure

tonated substrate. The energy difference between the two structures is very small. If the model is optimized in a solvent environment, a protonated lysine and a deprotonated substrate appear in the reactant, but there is no effect on the barrier compared to optimization in gas phase. The computed barrier for decarboxylation with this model is 37 kcal/mol, which is more than 20 kcal/mol higher than the barrier in the enzyme. The calculated barrier is in fact quite similar to the uncatalyzed reaction barrier. A single lysine thus has no catalytic effect on this reaction.

3.2 Effects of Hydrogen Bonds to O2 and O4

Starting with the small model, it is possible to make additions that can help stabilize the TS. As discussed above, a negative charge is transferred to the ring when the C–C bond is cleaved. This means that the two carbonyl oxygens O2 and O4 become more negative as the length of the C–C bond increases. Hydrogen bonds to these two oxygens will therefore increase in strength when going from reactant to TS and should therefore reduce the activation barrier. In three of the X-ray structures, O2 accepts a hydrogen bond from the invariant Gln215. O4 accepts a hydrogen bond from the backbone in all structures. A single hydrogen bond to O2 (Model 2 in Fig. 4) or O4 lowers the barrier, but only by 1 kcal/mol. In these calculations, formic acid is used as a general hydrogen bond donor. The most significant stabilization was obtained by forming a hydrogen bond chain between Lys93 and O2 (or O4), via one water molecule and a formic acid (Model 3 in Fig. 4). This configuration lowers the barrier by 4 kcal/mol for O2 and 6 kcal/mol for O4. However, the arrangement of amino acids in Models 2 and 3 does not correspond to the X-ray structure, and the 6 kcal/mol value should rather be taken as an upper limit for this hydrogen bond effect. It has been found experimentally that substituting O4 for sulfur decreases the reaction rate by 50% [2]. Assuming that this largely depends on loss of hydrogen bonding with sulfur instead of oxygen, this value corresponds to a relative change in hydrogen bond strength of 1 kcal/mol rather than 6 kcal/mol.

3.3 Including the Network of Charged and Invariant Amino Acids

Returning to the carboxylate side, one reason for the high barrier is the strong interaction between Lys93 and the carboxylate group in the reactant. As carbon dioxide is being formed, this interaction decreases in strength, which contributes to the high barrier. One way to decrease the attractive interaction between Lys93 and carboxylate in the reactant is to add the nearest negative aspartate (Asp91) to the model (Model 4 in Fig. 5). As mentioned in the introduction, Asp91 and Lys93 are both parts of a network of charged

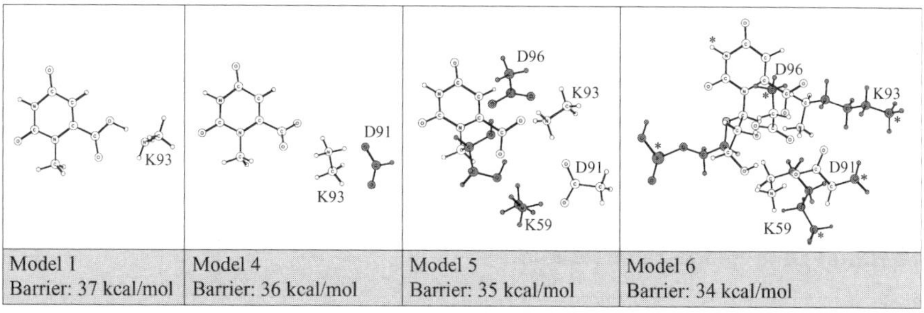

Fig. 5 QM models used for investigations of the concerted reaction mechanism where Lys93 protonates C6. This series of models shows the effects of residues close to the carboxylate. Extensions made to a previous model in the series are *shaded* to facilitate comparison between models. Note that despite the increase in the size of the models, the barrier is stable and high. Model 6 contains atoms that are kept frozen during the optimization and each of these atoms is marked with an *asterisk* (*)

invariant amino acids. Adding the aspartate leads to a weaker electrostatic attraction between the carboxylate and the lysine, as can be seen from an increase in the hydrogen bond distance from 1.4 Å to 1.7 Å. At the same time, the proton going to C6 is more reluctant to leave the lysine. In total, this leads to a very small effect on the reaction barrier (−1 kcal/mol).

As the invariant network of charged amino acids from mutation experiments is known to be very important for the enzymatic reaction [15], the computational models were extended with the other two invariant amino acids in this network, Lys59 and Asp96B. They were added stepwise in different configurations but at the end it was decided to keep the hydrogen bonding pattern from the X-ray structure. From the X-ray structure it is known that Lys59 and Asp96B are hydrogen bonded to 3′OH and 2′OH of the ribose ring, respectively. The substrate model was therefore extended to accommodate these details. If the structures are optimized in several steps with a gradual removal of the restrictions on the amino acids, it is possible to find local minima for reactant and TS that are structurally similar to the initial structure (Model 5 in Fig. 5). The hydrogen bond distance between Lys93 and carboxylate in the reactant is 1.65 Å, not far from the value in the Lys-Asp model (Model 4). The calculated barrier for this model including the entire network of charged amino acids is very high, 35 kcal/mol. Indeed, there is almost no difference between the result of this extended model and the simplest model with only one lysine (Model 1). In spite of the large variations between the models used, the calculated barrier is remarkably stable. Any effect on the interaction between the positive lysine and the carboxylate in the reactant seems to have a similar effect on the ability of the lysine to donate a proton in the TS. Lys93 is not the optimal residue for TS stabilization since it already interacts with two negative aspartates. An attractive solution would be a weak interaction between Lys93 and the carboxylate at the same time as other molecules are available to stabilize C6. In the X-ray structure, there are no other molecules available, but in another section of this review, structures with several weakly bound water molecules have been used in the calculations.

3.4
Effects of Long-Range Electrostatics

In the introduction it was stated that the eventual failure of a theoretical model to explain the catalytic effect of an enzyme could either be due to an incorrect reaction mechanism or to an incomplete model. To further investigate the latter case, attempts were made to look for possible effects outside the present model. How does the protein matrix influence the active site? One possibility is that an inhomogeneous charge distribution in the protein could have a large effect on the barrier since the charge distribution of the active site is different in reactant and TS. This type of electrostatic effect works over considerable distances since it decreases only slowly with distance. In order to investigate if charges outside the large QM model (Model 5) have a significant effect on the calculated barrier, all aspartate, glutamate, histidine, lysine, and arginine residues within 12 Å from the substrate in the yeast enzyme were added as point charges (Mulliken charge populations with d95 basis set) to the QM model. The point charges were placed at the same positions as they have relative to the phosphate group in the PDB-structure. Single point energies were then calculated for the reactant as well as for the TS and the influence of the point charges was noted. In what is considered as the most likely protonation state (Asp, Glu ,and phosphate negative, His neutral, Lys and Arg positive) the reaction barrier did decrease, but only by 3 kcal/mol. By varying the protonation states of the amino acids, effects on the reaction barrier from −3 to +4 kcal/mol were found. This is a rather primitive method and the results are not considered very reliable but they do indicate that these charges do not have any important effects on the barrier.

3.5
Analyzing the Influence of Protein Strain

Another possible effect of the protein matrix is to restrict the movements of the active site residues. Two different proposals for how ODCase achieves its catalytic power both require that the amino acid side chains are not completely free to move. In one QM/MM study, ground state destabilization was claimed to be the origin of the extreme catalytic activity of the ODCase enzyme [13]. This mechanism requires a strong repulsion between the negative carboxylate and Asp91 in the ground state (reactant). In the TS, this repulsion decreases as the carboxylate moves away from the active site. Another proposal for the enzymatic activity in ODCase includes the concept of pre-organization, which means that the enzyme should be aligned in a configuration that is already complementary to the TS. Before the reactant binds there should already be an electrostatic stress in the protein that is released when the structure reaches the TS [38]. Repulsion effects and electrostatic stress cannot exist unless the protein matrix keeps the active site in a rather firm grip. The results reviewed above originate from models where the residues are totally free to move. In those models, the calculated TS should have the lowest possible energy but the lack of strain may result in an artificially low energy of the reactant.

The implications of the electrostatic stress hypothesis were investigated separately using the Asp91-Lys93-Asp96B complex, which includes the amino acids found by to be the most important ones for catalysis [38]. The interaction energy of the amino acids was calculated quantum mechanically for the X-ray structure (1DQW) [12], and for the presently optimized TS and reactant structures for Model 5. Comparing the amino acid interaction energy in the optimized TS with that of the X-ray structure, the quantum mechanical calculations give a difference of only 2 kcal/mol, with a stronger interaction for the optimized TS. An interpretation of these results is that they support the hypothesis that the protein structure is optimized to stabilize the TS since the X-ray and optimized structures have similar interaction energies. On the other hand, comparing the interaction energy in the optimized reactant with that of the X-ray structure, the interaction energy of the optimized reactant is only 5 kcal/mol larger, indicating that the maximum electrostatic stress that could be introduced for the reactant is of this order of magnitude.

A very simple way to introduce strain from the protein in a way that potentially can include both ground state destabilization and electrostatic stress is to freeze atoms in the residue side chains. In this treatment, one atom in each side chain is frozen in its position in the X-ray structure. However, using this procedure for the large model (Model 5) described above is not possible since a rather large movement of the lysine is required. In the reactant it interacts with the carboxylate and in the TS it reaches all the way to C6. To allow for this movement, an increased part of the carbon side chain is included for each of the four amino acids (Model 6 in Fig. 5). Two atoms in the substrate are frozen to keep the substrate in the binding pocket. The carboxylate is then added, a procedure that brings it close to the enzyme residues. After optimization of the reactant, Lys93 and the carboxylate hardly interact at all. Lys93 instead forms a hydrogen bond to 2′OH. In the model with frozen atoms it is difficult to optimize a TS, therefore the C–C distance between C6 and the carboxylate was frozen at 2.88 Å and the C–H bond between C6 and Lys93 was frozen at 1.76 Å. These values were taken directly from the geometry of the fully optimized TS in Model 5. Despite the large additions to the model, the restrictions on the active site and the different binding in the reactant, there was no significant effect on the reaction barrier. It decreased by only 1 kcal/mol compared to Model 5 with completely free residues. If there still are any important effects of strain, this procedure did not show any sign of them.

3.6 Developing a QM/MM Treatment of the Concerted Mechanism

None of the investigations of the basic QM active site model (Model 5) indicate any major error in the treatment of this reaction. In spite of this, further extensions of the model were made. The next step was to extend the model with a more elaborate description of the surrounding protein, by including the protein outside the QM model in an MM description. A QM/MM treatment is expected to better treat the type of strain and long-range electrostatic effects that have been discussed. It should also be able to incorporate the idea that the TS induces a lower energy conformation of the protein than the reactant. This could lead to a lower barrier for the QM/MM system compared to the bare QM models.

No new search for a TS was made for the QM/MM models. Instead the energy of an optimized QM/MM structure with frozen a C–C bond distance of 2.88 Å (which corresponds to the TS geometry of Model 5) is compared to the fully optimized QM/MM reactant. QM/MM energies are obtained after a scan of the C–C distance from reactant to the assumed TS geometry in steps of 0.2–0.4 Å. This procedure is used to allow relaxation of the protein matrix after each minor change in the QM geometry. All QM/MM energies are calculated at the small basis set level and solvation effects are not included. The value of this relative energy between QM/MM reactant and TS is compared to the energy barrier for the corresponding QM model at its TS. From this comparison, the effect of the MM part is calculated. QM/MM barriers reported in Fig. 6 are derived from the free energy barriers of the QM models, adding the effect of the MM part.

Two QM/MM models with different sizes of the QM parts were used in this investigation. The small QM/MM model includes only Lys93 and the substrate in the QM part while the rest of the enzyme is treated with MM (Model 7 in Fig. 6). This model can be compared to the first QM model (Model 1) including the same residues as the QM part. A larger QM/MM model with substrate and the four charged amino acids treated as QM (Model 8) was designed for comparison to the corresponding QM model (Model 6).

First looking at the QM and QM/MM models with the small QM part (Models 1 and 7), an initial scan of the C–C distance in the QM/MM model resulted in a drop in the barrier height by over 10 kcal/mol. Analyzing the origin of this apparent TS stabilization, it appeared that the MM part of the

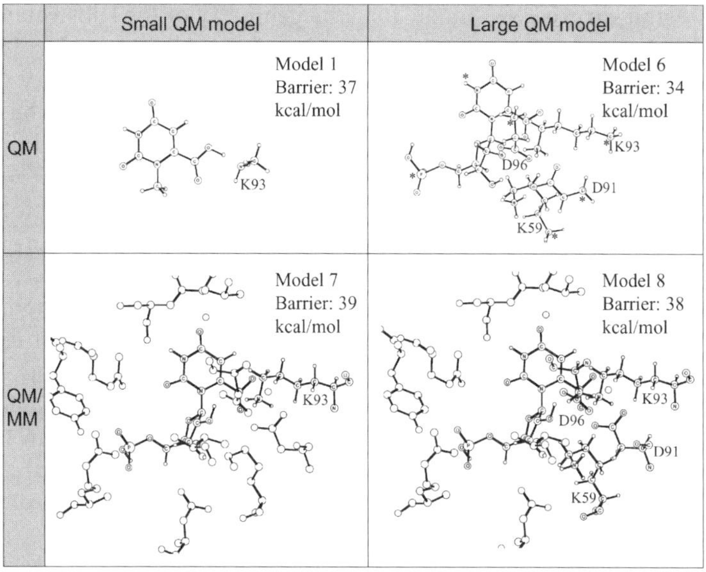

Fig. 6 Comparison of QM and QM/MM models used in the investigation of the concerted reaction mechanism where Lys93 protonates C6. For the QM/MM models, labeled atoms are included in the QM description. Hydrogen atoms in QM residues are not labeled and hydrogen atoms in MM residues are not displayed. The picture does not display all residues that are included in the MM part

enzyme outside the active site had found a new low energy configuration for the TS that had not been present in the reactant. In theory, this could be a real and important effect, but further investigations showed that it was an artifact. The low energy TS structure was allowed to relax and was followed back towards the reactant. This new reactant turned out to be 14 kcal/mol more stable than the first one and the apparent stabilization therefore disappeared if the TS was compared to the new reactant. This procedure was repeated until there was almost no change in energy between two consecutively optimized reactants. After four scans of the C–C distance and a difference between the fourth and the fifth reactant of 0.7 kcal/mol the procedure was stopped. The major structural difference between the QM model and the QM part of the QM/MM model is the interaction between Lys93 and the carboxylate. In the optimized QM/MM reactant the hydrogen bond distance between lysine and the carboxylate is 2.1 Å, which is a rather long distance compared to 1.4 Å in the model including lysine only, but this did not have any significant effect on the barrier. After the iterative procedure was completed, the apparent lowering of the barrier when the enzyme was included as MM disappeared. The final barrier, if compared to the energy of the TS of Model 1, increased by 2 kcal/mol, which is a quite small effect. It can even be considered surprisingly small since the geometry of the substrate and the active site changed significantly compared to the small QM model. The QM/MM model also included several charged amino acids located very close to the QM part (e.g., Lys59, Asp91, and Asp96B). However, it should be kept in mind that these amino acids did not have any large effect when going from a small to a large QM model (see Fig. 5).

The large QM/MM model (Model 8) includes over 100 QM atoms and more than 4,600 MM atoms but, despite the size of the model, there was no sign of any important enzymatic effects. The barrier actually increased by 4 kcal/mol when going from the large QM to the large QM/MM model. This QM/MM treatment should be able to describe several of the most common suggestions for how the protein matrix influences the reaction barrier. Among the effects included are electrostatic effects, strain effects, and different conformations in reactant and TS for the parts of the protein that are not included in the QM model. Although the MM energies are not as accurate as the QM energies, it would be an unlikely coincidence if errors in the QM/MM treatment masked a real protein effect.

The results presented do not indicate any urgent need to use QM/MM instead of QM models for this reaction with this QM model. In general, a QM/MM treatment is preferable to a QM treatment, but only if it is carefully performed and does not introduce erroneous effects. At the same time it involves a considerable effort to perform the iterations over different MM minima. If a significant effect is expected from the surrounding protein residues it can even be more efficient and accurate to extend the QM model to incorporate this effect.

A possible source of a too-high barrier is that the truncated QM model lacks molecules that are important for TS stabilization. A possibility is that weakly bound molecules, e.g., water molecules, enter the active site during the reaction and stabilize the TS. This suggestion cannot be investigated unless the water molecules are included in the QM model. However, the QM/MM model includes all water molecules that are present in the X-ray structure but the possibility of TS stabilization does not show up in the QM/MM calculations, probably because the optimizations only identify the minimum closest to the X-ray structure. Other QM/MM models that include dynamics (solvent equilibration) may be able to show these effects, as seen in the section describing the reaction mechanism where the C–C bond is cleaved prior to decarboxylation. Dynamics opens up new possibilities for reaction mechanisms by identifying mobile residues or molecules. This particular study cannot definitively judge the importance of such treatments. Perhaps it might still be possible to identify these possibilities and add the necessary elements to the QM model.

In summary, this section shows that despite the use of many different models for the concerted protonation mechanism, the calculated reaction barrier for this mechanism is always considerably higher than the experimental barrier. The geometry of the active site changes considerably, exemplified by the hydrogen bond distance between Lys93 and carboxylate, but these geometric changes do not have any major effect on the barrier. The main problem seems to be that either the lysine interacts too strongly with the carboxylate or it is too reluctant to donate a proton to C6. Stabilization of the negative charge developing on C6 must apparently use other means than Lys93.

4 Base Protonation Mechanisms

Early proposals for the reaction mechanism of ODCase favored mechanisms where the pyrimidine ring is protonated prior to decarboxylation [5, 6]. Theoretical investigations have shown that a protonation of the pyrimidine ring leads to a significant decrease in the strength of the C–C bond [39]. Furthermore, kinetic studies indicating that the enzyme kinetics are governed by two ionizations [17, 40] can be interpreted to favor a base protonation mechanism. The first ionization could be the protonation of the base, and the second one is likely to be the final proton donation to C6. On the basis of another kinetic study, Rishavy et al. claimed that base protonation mechanisms involving O2 or O4 do not match the measured kinetic isotope effects [41], but later quantum chemical calculations showed that both base protonation mechanisms and mechanisms without protonation are consistent with the measured isotope effects [42]. From the X-ray structure it is clear that there are no obvious proton donors close to any of the protonation sites. This has been interpreted as evidence against a protonation mechanism, but the only actual structural requirement is that there exists a possible proton pathway to reach the protonation site.

For the base protonation mechanism there are two major questions. The first question is which site of the pyrimidine ring should be protonated. Over the years, several suggestions have been made that favor specific protonation sites [2, 5, 6, 7, 21, 39]. Experimental results seemed to favor O2 protonation but at present the collected evidence seems inconclusive [2, 43, 44]. Since the strength of the C–C bond depends only weakly on the particular site of protonation [39], it appears that the decisive factor should be the relative proton affinities of the different sites and this property is the first to be investigated below. The second major question for the base protonation mechanism concerns the energetics of the C–C bond cleavage. To estimate the total barrier for the base protonation mechanism both the barrier for C–C bond cleavage and the cost of protonating the base must be calculated. If protonation is not spontaneous, the protonation cost must be added to the energy required to cleave the C–C bond. The total barrier for these two steps should be around 15 kcal/mol to match the experimental reaction barrier.

4.1 Calculating Protonation Costs Using QM Models

The main challenge in the present type of quantum mechanical modeling is to estimate the protonation cost. The proton needed for the substrate reaction is ultimately provided by the solvent, a part that cannot be included in the model. To be able to work with a limited model, it is assumed that the resting state of the proton is the position of lowest energy in the quantum chemical model. For most models this position turns out to be the carboxylate. This does not mean that the proton actually comes from the carboxylate or that the mechanism requires that the carboxylate is protonated in the reactant. The procedure simply gives a lower limit for the energy required to protonate the base.

To compare the proton affinities of different sites of the pyrimidine ring a naked small substrate model (Model 9) was first used, for which earlier calculations had already shown that O4 has a higher proton affinity than O2 [39]. Using this model, the proton affinities of all possible sites of the pyrimidine ring were investigated. The results below are given in terms of the cost to move a proton from the carboxylate group, rather than reporting absolute proton affinities. The calculations confirmed that in the small substrate model the carboxylate is the best site to protonate, followed by the O4 site with an energy level 15 kcal/mol higher. In this model, C5 and O2 are much more costly to protonate (27 and 28 kcal/mol, respectively), while N3 is the worst candidate (cost of 42 kcal/mol). In their study, Kollman et al. found that for a substrate model, the proton affinity of C5 is similar to the proton affinity of O4 [7]. The origin of the different results for protonation of C5 is not clear but a possibility may be a different treatment of solvation effects, which are rather large for these models.

After the first investigation, the model was extended to include the full substrate and a lysine at the carboxylate side (Model 10 in Fig. 7). In this model, the carboxylate decreases its proton affinity due to the positive lysine. At the same time O2 increases its proton affinity, both because the proton on O2 can form a hydrogen bond to the ribose ring, and because the negative phosphate attracts the O2 proton. In this model, the O2 protonation cost relative to the carboxylate is down to 11 kcal/mol, which is approximately the same value as for O4 in this model. A further extension of the model is to include all residues that form hydrogen bonds with the pyrimidine ring (Model 11 in Fig. 7). Amino acids that are hydrogen bond donors to a protonation site will decrease the proton affinity of that site and these are Gln215 to O2, a backbone amide proton from Ser154 to O4, and water to the carboxylate. Amino acids that accept hydrogen bonds increase the proton affinity of the corresponding site. In this active site there is only Ser154 that accepts a hydrogen bond from N3. C5 is surrounded by hydrophobic residues that are not included. In Model 11 the decrease of the proton affini-

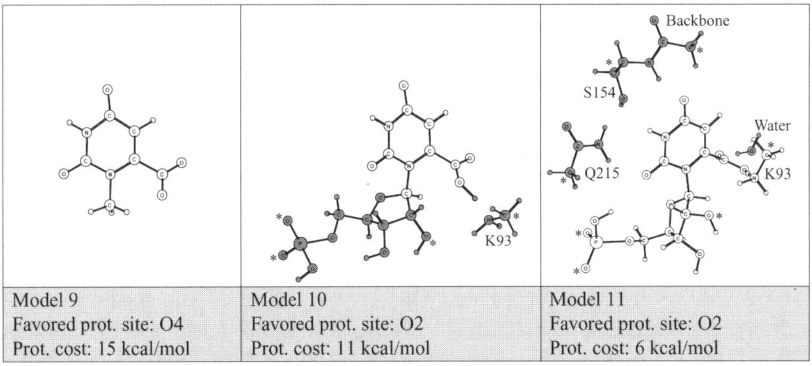

Fig. 7 QM models used to calculate the proton affinity of all protonation sites in the pyrimidine ring. This series of models shows the effects of extensions to the substrate model and of residues hydrogen bonding to the ring. Extensions made to a previous model in the series are shaded to facilitate comparison between models. Models 10 and 11 contain atoms that are kept frozen during the optimization and each of these atoms are marked with an *asterisk* (*)

ty of O2 is smaller than the effect on the other sites, which makes O2 the best protonation site, after the carboxylate, with a cost of only 6 kcal/mol relative to the carboxylate. This is 7 kcal/mol lower than the cost of protonating O4 and 15 kcal/mol lower than the cost of protonating C5.

As the carboxylate is the site with the highest proton affinity, it was first investigated whether protonation of this site could actually lead to a low barrier for the concerted mechanism discussed in the previous section. Harris et al. suggested that such a proton would be located between the carboxylate and Asp91, and that this could possibly assist in catalysis [14]. It seems reasonable that an extra proton in this region would stabilize the negative charge developing on C6, and therefore a proton was added to Model 4 that includes Lys93 and Asp91. The addition of the proton turned out to increase the barrier for decarboxylation by 8 kcal/mol in this model. The increase of the barrier can be interpreted as the cost to break two strong hydrogen bonds in the reactant rather than one in the original reactant, the one between the carboxylate and the protonated Asp91, and the one between the carboxylate and Lys93. The conclusion is that the proton must be located at some other place in the active site, if it should lower the barrier for decarboxylation.

The above results thus indicate that O2 is the most favorable protonation site after the carboxylate. In the following investigations of the C–C bond cleavage barrier, the O2 site is assumed to be protonated. It could be argued that including the negative groups Asp91 and Asp96B would probably increase the proton affinities of O4 and C5 closer to the value for O2, since these protonation sites are closer to the negative residues. However, this effect of the model extension would also increase the cost of protonating these sites relative to the carboxylate so there seems to be little point in investigating this possibility.

4.2
Calculating Reaction Barriers Using Improved Models of the Charged Network

What remains to be found is how improved models of the region around the carboxylate will affect the strength of the C–C bond and if the proton affinity of O2 can be further increased. The investigation of the barrier for C–C bond cleavage with a protonated substrate started with a bare substrate model (Model 9 in Fig. 8) used by Singleton et al. [39]. The strength of the C–C bond is taken as the highest point on the free energy curve for the C–C bond stretch. The electronic energy increases monotonically with the distance, but by adding the entropy contribution for each distance, a free energy barrier of 20 kcal/mol is obtained for this model, at a C–C distance of 2.4 Å. When going to larger models, this C–C distance will be used to probe the barrier height for the C–C bond cleavage. The calculated bond strength of 20 kcal/mol for the protonated substrate (Model 9) is significantly lower than for the unprotonated substrate (by approx. 20 kcal/mol), but if the cost for the O2 protonation, 28 kcal/mol for this model, is added, the total barrier becomes as large as 48 kcal/mol. An interesting question is to what extent this result can be modified by the introduction of enzyme residues in the quantum mechanical model. The same series of models as were used for the concerted mechanisms will initially be discussed.

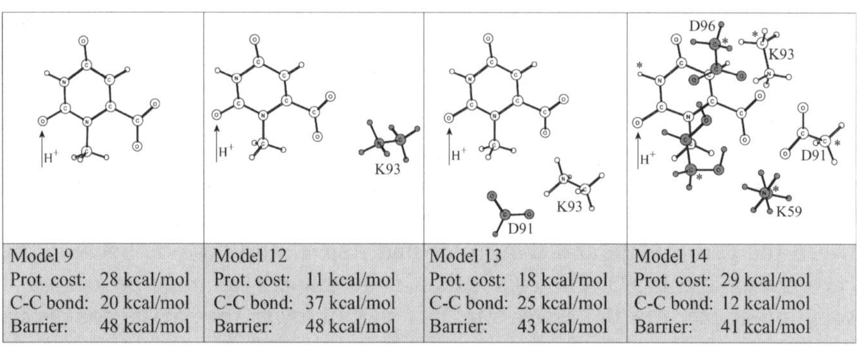

Model 9	Model 12	Model 13	Model 14
Prot. cost: 28 kcal/mol	Prot. cost: 11 kcal/mol	Prot. cost: 18 kcal/mol	Prot. cost: 29 kcal/mol
C-C bond: 20 kcal/mol	C-C bond: 37 kcal/mol	C-C bond: 25 kcal/mol	C-C bond: 12 kcal/mol
Barrier: 48 kcal/mol	Barrier: 48 kcal/mol	Barrier: 43 kcal/mol	Barrier: 41 kcal/mol

Fig. 8 QM models used to calculate the energy required to protonate O2 and the strength of the C–C bond. The reaction barrier is calculated at a frozen C–C distance of 2.4 Å. This series of models shows the effects of residues close to the carboxylate. Extensions made to a previous model in the series are *shaded* to facilitate comparison between models. Model 14 contains atoms that are kept frozen during the optimization and each of these atoms is marked with an *asterisk* (*)

The first extension of the model is to add Lys93, which interacts directly with the carboxylate (Model 12 in Fig. 8). This actually leads to an increase in the C–C bond energy by 17 kcal/mol, giving a C–C bond cleavage barrier of 37 kcal/mol. In this model the strong interaction between lysine and carboxylate in the reactant, discussed already in the section on the concerted mechanism, is directly added to the barrier. When the carboxylate is on its way to becoming carbon dioxide, this electrostatic interaction is drastically reduced. However, at the same time the lysine decreases the proton affinity of the carboxylate, and thereby the cost to protonate O2 is decreased to 11 kcal/mol, if the proton affinity of the carboxylate is used as reference. The total barrier for decarboxylation in this model is thus 48 kcal/mol, the same value as for the naked substrate model.

It is clear that if charged amino acid residues are included in the model, and if there are oppositely charged residues in the vicinity, those should be included as well. The next step is therefore to also add the nearest aspartate to the model, which should decrease the harmful electrostatic attraction between the carboxylate and the lysine in the reactant (Model 13 in Fig. 8). For this model the energy required to cleave the C–C bond decreases to 25 kcal/mol, and the cost to protonate O2 increases to 18 kcal/mol, giving a total barrier of 43 kcal/mol. The next extension of the model is to include all four charged amino acids close to the carboxylate (Lys59, Asp91, Lys93, and Asp96B) with the hydrogen bonding configuration taken from the X-ray structure (Model 14 in Fig. 8). With this configuration, the barrier for decarboxylation decreases significantly and becomes 12 kcal/mol. Compared to the bare substrate model (Model 9) with a barrier of 20 kcal/mol, there is thus a significant catalytic effect on the C–C bond cleavage barrier from the charged network of 8 kcal/mol. This is a much larger effect than was obtained for the concerted decarboxylation mechanism, see Fig. 5. To obtain this effect a specific configuration of the network of charged amino acids is required. These comparisons give a nice illustration of how the influence from particular enzyme residues can be quantified in this type of modeling. It must be noted, however, that the O2 protonation cost for this model is as large as 29 kcal/mol, yielding a total decarboxylation barrier of 41 kcal/mol.

4.3 Calculating Reaction Barriers Using Improved Models of the Environment around O2

So far all models used for the base protonation mechanism have given total barriers for decarboxylation of more than 40 kcal/mol. In all those models, the amino acids added were chosen mainly to affect the C–C bond strength, but a large fraction of the total barrier comes from the protonation cost. The next step is therefore to discuss models that focus on the region around O2 and include residues that might affect the proton affinity of O2. In the X-ray structures only the invariant Gln215 interacts with O2. In most X-ray structures, this residue donates a hydrogen bond to O2, which should decrease the proton affinity of O2. However, in one of the structures the interaction between Gln215 and O2 is mediated by a water molecule [13]. One possibility is that in this structure, Gln215 instead accepts a hydrogen bond to its carbonyl oxygen, mediated by the water molecule. This model ignores the possibility that water and Gln215 already have several other hydrogen bonds in the reactant and the proton affinity of O2 in this model can be taken as an upper limit. If this configuration is used in the model (Model 15 in Fig. 9) the cost to move the proton from the carboxylate to O2 is reduced by 12 kcal/mol compared to the naked substrate (Model 9). The stabilization of the proton at O2 is due to the formation of a chain of strong hydrogen bonds from the proton at O2 to the carbonyl oxygen on Gln215 mediated by the water molecule. The C–C bond cleavage barrier is unchanged in this model, leading to a total barrier for decarboxylation of 36 kcal/mol. This increase of the model should be combined with the models where the carboxylate surrounding is included, and therefore two models were designed with residues on both sides of the substrate. The first of these models has only a lysine on the carboxylate side, and a glutamine and water on the O2 side (Model 16). In this model, protonation of O2 is actually favored compared to the carboxylate, by 5 kcal/mol. However, from Model 12 (Fig. 8) it is known that this model has a very strong C–C bond. Therefore, adding both the protonation step and the C–C bond strength gives a decarboxylation barrier of 38 kcal/mol.

The final combined model should be the most balanced one, and it adds the O2 residues to Model 14, which has the charged network on the carboxylate side. In this model (Model 17), O2 protonation costs only 7 kcal/mol. This is surprisingly low, since by comparing the results for the Models 14

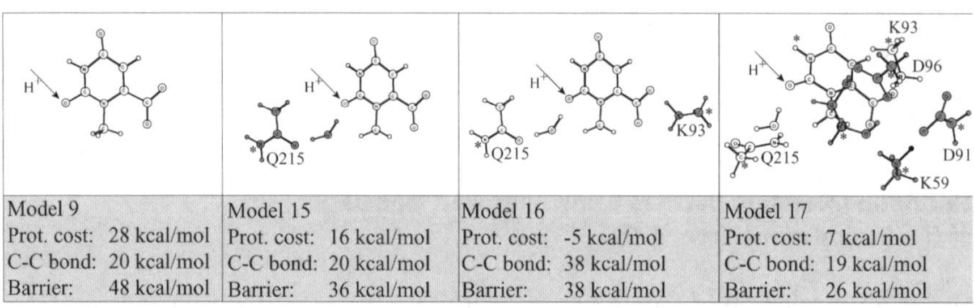

Model 9	Model 15	Model 16	Model 17
Prot. cost: 28 kcal/mol	Prot. cost: 16 kcal/mol	Prot. cost: -5 kcal/mol	Prot. cost: 7 kcal/mol
C-C bond: 20 kcal/mol	C-C bond: 20 kcal/mol	C-C bond: 38 kcal/mol	C-C bond: 19 kcal/mol
Barrier: 48 kcal/mol	Barrier: 36 kcal/mol	Barrier: 38 kcal/mol	Barrier: 26 kcal/mol

Fig. 9 QM models used to calculate the energy required to protonate O2 and the strength of the C–C bond. The reaction barrier is calculated at a frozen C–C distance of 2.4 Å. This series of models shows the effects of residues close to O2 and the carboxylate. Extensions made to a previous model in the series are *shaded* to facilitate comparison between models. Models 15–17 contain atoms that are kept frozen during the optimization and each of these atoms is marked with an *asterisk* (*)

(Fig. 8) and 15 (Fig. 9) a value of 17 kcal/mol is expected, if the effects from the two sides were additive. On the other hand, the C–C bond cleavage barrier has a corresponding unexpected effect in the opposite direction, with a value of 19 kcal/mol. In fact this means that for the total decarboxylation barrier, the effects from adding residues on both sides are essentially additive, and the total barrier becomes 26 kcal/mol. Thus the inclusion of the amino acid residues in the model has brought down the barrier for decarboxylation according to the base protonation mechanism from 48 kcal/mol in the naked substrate model (Model 9) to 26 kcal/mol in Model 17, indicating a catalytic effect of the enzyme of 22 kcal/mol, close to the experimental value of 23 kcal/mol. Still, the height of the calculated barrier for O2 protonation in the enzyme (26 kcal/mol) appears to be too far from the experimental barrier of 15 kcal/mol to support the base protonation mechanism.

In an attempt to bring down the calculated barrier for O2 protonation further below the value of 26 kcal/mol, a final model of the O2 side was designed (Model 18 in Fig. 10). The phosphate, and groups interacting with the phosphate, have been shown to have important effects on the rate of catalysis (k_{cat}) [45, 46]. Essentially the same structure that gave the low protonation barriers (Model 11) is used, but here more hydrogen bonding possibilities are present. If a singly protonated phosphate is used as reference state instead of the carboxylate, the energy required to protonate O2 is as high as 32 kcal/mol. In this extended model, Gln215 does not stabilize the proton at O2. This indicates that the protonation cost of O2 calculated for Models 15–17 could be underestimated.

The above results, using extended quantum mechanical models for the active site of ODCase, indicate that the base protonation mechanism has a decarboxylation barrier that is too high to be compatible with the experi-

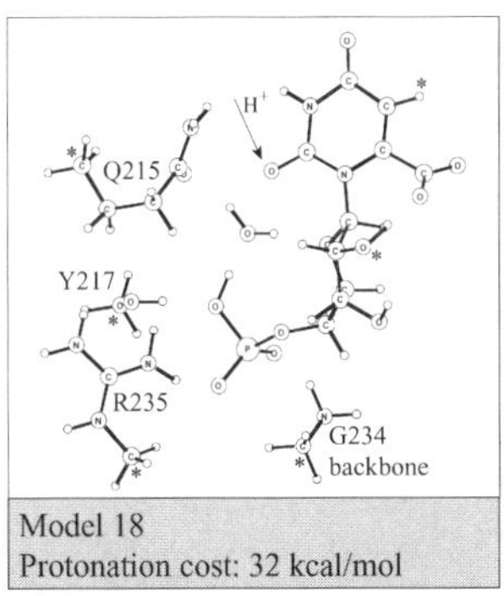

Fig. 10 QM model used to calculate the energy required to protonate O2. In this reaction, the reference state is a proton at the phosphate group. This model contains atoms that are kept frozen during the optimization and each of these atoms is marked with an *asterisk* (*)

mental rate of the enzyme. It should be noted, however, that to calculate the barrier for that mechanism, proton affinities of different sites have to be compared. As is also clear from the results above, proton affinities are extremely sensitive to the environment, and it could be argued that the models used above are still too small to give a balanced description of the protonation of different sites. This is apparent in the modeling of O2 protonation where the orientations of water and Gln215 have a large influence on the protonation cost. In one model, the cost to protonate O2 is significantly lower, 7 kcal/mol, but further extensions of the model around O2 indicate that this value might be too low. Using 7 kcal/mol as protonation cost, the total reaction barrier is still 26 kcal/mol. For the protonation sites O4 and C5, the effect of the charged network at the carboxylate side (Model 14) should be added to the effect of residues hydrogen bonding to the protonation sites (Model 11). With this procedure, the cost to protonate O4 and C5 is around 25 kcal/mol, while the cost to protonate N3 is even higher.

4.4 Calculating Protonation Costs for the Pyrimidine Ring Using QM/MM Models

An alternative to adding effects from different models is to design a large QM/MM model that includes all residues. Two models, differing in the size of the QM model, were used to calculate the proton affinity of all different sites in the pyrimidine ring (Models 19 and 20 in Fig. 11). The results from these models affirm the conclusion drawn from the QM models that if the cost of protonation is added to the C–C bond strength, the reaction barrier is too high. A problem with the QM/MM models is that proton affinities for sites that interact with MM residues are overestimated. When changing the amino acids Leu153, Ser154, and Gln215 from MM to QM (i.e., going from

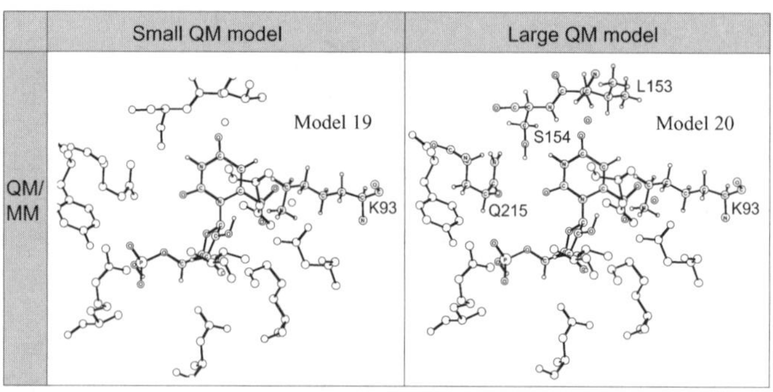

Fig. 11 QM/MM models used to calculate the proton affinity of all protonation sites in the pyrimidine ring. Labeled atoms are included in the QM description. Hydrogen atoms in QM residues are not labeled and hydrogen atoms in MM residues are not displayed. The picture does not display all residues that are included in the MM part

the small to the large QM/MM model) the calculated proton affinities go down by an average of 17 kcal/mol. There is a similar effect for the carboxylate but it shows up in both the small and the large QM/MM models since the charged network (except Lys93) in both models is treated with MM. The effect on the carboxylate was confirmed if the QM Model 14 (Fig. 8) was treated with QM/MM, having all protein residues but Lys93 in the MM part. It is important to be aware of this problem since small QM/MM models may underestimate the cost of protonating the substrate.

Summarizing the treatment of the base protonation mechanism, it has been shown that a certain configuration of the charged network is required for cleaving the C–C bond with a reasonable barrier. This configuration exactly mimics the one found in the X-ray structures of the enzyme. It is therefore likely that the enzymatic mechanism uses the charged network in the way outlined for the base protonation mechanism. Still, the base protonation mechanism cannot be supported in full. The problem with the base protonation mechanism is that the proton affinities of all sites appear to be too low. If the cost of protonation is added to the C–C bond strength, the reaction barrier is too high.

5 C–C Bond Cleavage Prior to Proton Donation

The conclusion from the two previous sections is that the transition state stabilization provided by protonation of the base or by the Lys93 proton is not enough to give a low barrier for the decarboxylation reaction. An interesting alternative is the stepwise mechanism proposed by Warshel et al., in which the C–C bond is cleaved prior to protonation of C6 and where the transition state stabilization is at least partly provided by water molecules close to C6 [38]. In their study, the catalytic reaction was described using an empirical valence bond potential energy surface, calibrated against quantum mechanical calculations on the reaction in solution. Using this potential energy surface for the active site, and performing molecular dynamics simulations on the enzyme, a reaction barrier of 17–24 kcal/mol was found, where the precise value depends on the protonation states chosen for the nearby amino acids. It was argued that the origin of the enzymatic activity of ODCase was mainly an effect of pre-organization, which means that the enzyme from the beginning is aligned in a configuration that fits the transition state structure of the substrate. When the reactant binds there is then an electrostatic stress in the protein, which is released when the substrate reaches the transition state structure. These ideas are described in detail in references [47, 48, 49].

5.1
Reactant and Intermediate with a Restricted QM Model

In a previous paper the results obtained by Warshel et al. were analyzed using limited QM active site models, and the results from that study indicated that the pre-organization energy could not provide the necessary TS stabilization [21]. Recently, the structures resulting from the study of Warshel et al. have been used as starting points for the type of QM modeling presented in the present review. Two structures were provided to us from the authors, one reactant structure, W-R, and one structure where the C–C bond is cleaved, W-I, corresponding to an intermediate for their mechanism. Those structures had been obtained by independent molecular dynamics simulations for the reactant and TS regions, which means that there can be significantly different configurations of the rest of the enzyme in these structures. It could seem as if such a procedure would depend strongly on an accurate description of the relative energies of different enzyme structures, but it should be noted that the relative free energies are obtained by averaging over many different configurations. This is a rather different approach than the procedure followed in the QM modeling described above, where the enzyme is kept in as similar configurations as possible for reactant and TS.

The structures W-R and W-I differ significantly in the active site region from the X-ray structure of the yeast enzyme that was used in the QM modeling described in the previous sections. An important aspect is that there are three water molecules located close to the carboxylate, rather than one as in the X-ray structure. The invariant network of four charged amino acids is connected to Asp37, which forms a hydrogen bond to Lys59. Furthermore, Asp96B no longer hydrogen bonds to 2′OH of the ribose ring. In total this gives a hydrogen bonded network with a larger radius, which on average is further away from the substrate than in the X-ray structure.

The QM models extracted from the structures W-R and W-I contain the four charged amino acids Lys59, Asp91, Lys 93, and Asp96B as in previous models, but now the three water molecules in the active site are also included (Model 21). The reactant model and the intermediate state model are thus constructed from different structures as described above, with the purpose of computing the relative energy between them within the QM model. Entropy effects, which are not included here, are expected to lower the intermediate where the C–C bond is cleaved by a few kcal/mol, and therefore the computed relative energy should be corrected by this energy. Initial bond distances and angles in the model constructed are far from optimal in a QM description, and therefore these have to be optimized before an energetic comparison can be done. To keep the overall geometries close to the starting structures, several atoms were kept frozen during the optimizations as indicated in Fig. 12. After optimization of the two structures, a relative energy of only 15 kcal/mol was obtained, thermal and entropy effects not included.

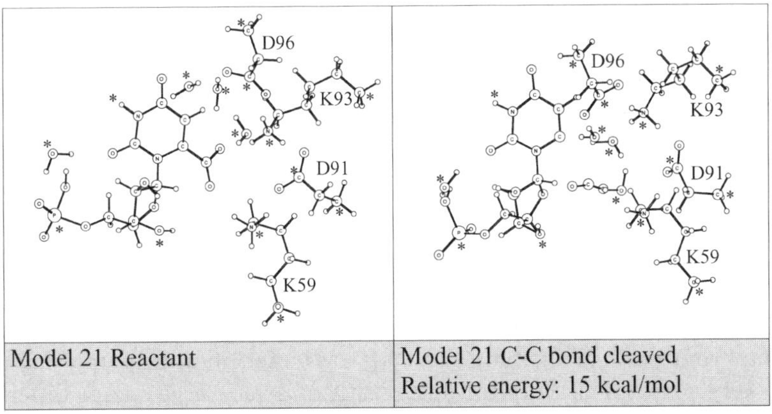

Fig. 12 Reactant and intermediate structures for investigation of a stepwise mechanism where the C–C bond is cleaved prior to protonation of C6. In the QM optimization of the structures, the atoms marked with *asterisks* (*) were kept frozen, from the structures obtained by molecular dynamics simulations. The relative energy reported does not include thermal or entropy effects

This is a remarkable decrease from the high barrier of the concerted mechanism. The geometry of the intermediate state, with water molecules interacting with C6, indicates that the water molecules are important, and in this mechanism they seem to provide an essential part of the stabilization of the negative charge developing on C6. As mentioned above, these water molecules were not included in the previous QM modeling studies. The way the invariant network of charged amino acids is used in this mechanism resembles the configuration that provided TS stabilization in the base protonation mechanism (Model 14 in Fig. 8).

Comparing these two mechanisms, the origin of the stabilization of the negative charge has changed from a proton on O2 to the weakly bound water molecules that find their way to C6 of the substrate. The increase in vacant space between the amino acid network and C6 of the substrate, which is the result of the molecular dynamics treatment, might be necessary to allow for the water molecules to approach the pyrimidine ring. However, the electrostatic effects are well described by the rather small QM model and there does not seem to be any need to describe any long-range electrostatic effects.

The results described above, yielding a relative energy of only 15 kcal/mol between the reactant and the intermediate state, are not fully conclusive, since the structures in Fig. 12 are too severely constrained to be considered as good models of the active site. However, if the structures are released, there are large movements of the protein residues and the water molecules, and the stabilization disappears. Either the constrained structures are too far from true minima or the QM model lacks important parts that force residues and water molecules to stay in the right conformation. The latter alternative can only be analyzed by extending the computational model, and since the QM model already consists of 107 atoms, the only possibility is to try a QM/MM treatment.

5.2 Reactant and TS with a QM/MM Model

The two structures W-R and W-I were used as starting points in a QM/MM treatment. In a similar way as for the QM modeling above, the coordinates are far from optimal in the Qsite QM/MM description and optimizations have to be performed. After the optimizations, a very large energy difference (120 kcal/mol) remains between reactant and intermediate. Since the original structures were so far from any QM/MM minimum, it is possible that the optimizations resulted in artificial conformations, which do not describe the true difference between reactant and TS. The present conclusion is that it is very hard to obtain realistic relative energies by performing QM/MM calculations starting from the two different structures.

Instead, the regular procedure of directly connecting the reactant structure with a TS structure having a similar conformation of the protein was applied. Starting from W-R, the C–C bond was increased in several steps of QM/MM calculations, iterating back and forth between reactant and transition state to find a similar energy for two consecutive reactants. After four iterations this procedure had still not converged but the energy required to extend the C–C bond to 2.4 Å was rather stable at 22–25 kcal/mol. In this structure, two water molecules had found their way to C6 exactly as in the W-I structure. The results include the effect of the total MM energy of the two configurations as well as the effect of the MM part on the QM part. However, since the iterative procedure produce two configurations that are similar, the MM energies tend to be very similar for reactant and TS. Adding thermal corrections from other calculations on the decarboxylation reaction, the calculated relative energy corresponds to a barrier around 19–22 kcal/mol, still not including solvent effects. Although this result does not perfectly match the experimental barrier of 15 kcal/mol, there is a significant catalytic effect. The result is in reasonable agreement with the 17–24 kcal/mol obtained by Warshel et al. in their study of ODCase [38]. At this stage of the investigation, this stepwise mechanism looks like the most promising one.

5.3 Reproducing the QM/MM Results with a QM Model

Finally, an attempt was made to design a QM model that would show the same energetics for C–C bond cleavage as the above QM/MM calculations did. This model includes the five charged amino acids Asp37, Lys59, Asp91, Lys93, Asp96B, and the three water molecules at the active site. Starting from the most stable QM/MM reactant geometry the structure was optimized in QM. Attempts were then made to start from this structure and search for a low energy TS in the QM model, but these attempts failed. The amino acid residues and the water molecules did not find the transition state positions they did in the QM/MM model, and which gave a reasonably low barrier.

Although the investigation was in no way exhaustive, it seems like a QM/MM treatment is advantageous in this case. In a QM/MM description all hydrogen bonds are saturated and the positions of the mobile water molecules are much better defined. If there are a large number of unsaturated hydrogen bonds in the QM model, artificial changes in hydrogen bonding patterns may destroy the computational results. This is well known, but in this mechanism it seems to be unusually critical due to the large amount of unbound water molecules and mobile charged residues. The molecular dynamics treatment also provided some new features that were not discovered in a pure QM modeling. For example, those calculations gave the possibility to find reactant and TS structures that did not correspond exactly to the X-ray structure. An important feature is the appearance of three water molecules in the active site, which change their positions significantly between reactant and TS. In the X-ray structure used as starting point for all QM modeling, except the one described above, there is only one water molecule close to the carboxylate. This water molecule was not included in the QM models of the concerted protonation mechanism, and in the QM/MM models it does not move to C6 during decarboxylation.

6
Insights/Discussion

The computational results presented in this review give a clear picture regarding the concerted protonation mechanism for decarboxylation. For this mechanism, where C6 is protonated by Lys93 simultaneously with the C–C bond cleavage, the calculated barrier is simply too high compared to the experimental rate. A large variety of models have been used, ranging from small QM models to large QM/MM models, and the results for the barrier are quite stable at around 36 kcal/mol. The conclusion is that the energetics of this mechanism are already well described by the smallest model, involving only the substrate and one protein residue, Lys93, and the inclusion of the surrounding enzyme does not affect the barrier. This does not mean that the enzyme is just a spectator; it only means that the energetic effects from the enzyme are very similar for the reactant and the transition state of the concerted mechanism. This is illustrated by the distance between Lys93 and the substrate carboxylate, which changes from 1.4 Å in a small model including only Lys93 (Model 1) to 2.1 Å in a larger model including the entire enzyme in a QM/MM description (Model 8). Despite this large change in hydrogen bond distance, the reaction barrier for decarboxylation only changed by 1 kcal/mol. Models that give weak interactions between Lys93 and the substrate carboxylate in the reactant also make Lys93 a weak proton donor in the transition state, and vice versa. The failure to find any enzymatic effect when including parts of the enzyme in the computational model shows that the concerted mechanism is not the one employed by the enzyme. In general, there is of course a large difference in reaction barrier when modeling the substrate reaction with or without enzyme. However, if the mechanism is correctly described, relatively small QM models can catch the important electrostatic effects, assuming that the models are properly selected.

It has been suggested that the ODCase enzyme works through a ground state destabilization mechanism, in which electrostatic repulsion between the substrate carboxylate and Asp91 should raise the energy of the reactant compared to the transition state. No such effect could be found from the modeling performed. It could be argued that the QM models used have too little strain from the surrounding protein to describe such an effect, but if there had been any significant repulsion between Asp91 and the substrate carboxylate this would have appeared in the QM/MM calculations. Repulsion between these two negative groups would also have resulted in a significant increase in the proton affinity of the carboxylate, but no such effect could be found. When going from the bare substrate model (Model 9) to a model including the network of charged amino acids (Model 14), the proton affinity of the carboxylate did not change. This indicates that the nearby Lys93 effectively cancels any repulsion. Other arguments against ground state destabilization have been put forward by other groups [38, 50, 51].

For the base protonation mechanisms, i.e., where the pyrimidine base is protonated prior to decarboxylation, the calculations also indicate that the barrier is too high to be consistent with the rate of the enzyme. In this case the energetics of two different steps have to be summarized to give the final decarboxylation barrier. The first step is the cost of protonating the base, and the second step is the actual barrier for C–C bond cleavage when the base is protonated. In contrast to the picture for the concerted mechanism, the calculations show that both steps of the base protonation mechanism are quite sensitive to the model used. This model sensitivity is particularly pronounced for the first step, the protonation cost, which is an essential part of the total barrier. Since the immediate source of the proton is not known, the protonation cost is calculated relative to the carboxylate group, which has the largest proton affinity in most models used. Thus, relative proton affinities have been calculated, and the protonation cost is found to vary between −5 and +29 kcal/mol, depending on the model. For some of those results it is obvious that the model used is unbalanced between different sites, but it is clear that the QM model that is required to obtain accurate estimates of the proton affinity on all protonation sites simultaneously would be too large to be handled effectively. The barrier for the entire decarboxylation reaction according to this mechanism is somewhat less sensitive to the model than the individual steps. The lowest total barrier obtained is 26 kcal/mol, but comparisons of proton affinities between different models, all treating the environment around O2, indicate that this value might be an underestimation of the true barrier. Therefore, it is concluded that this type of mechanism is not very likely.

An option when encountering a need for very large models is to include a part of the system in a MM description. The results obtained when trying to do this for the relative proton affinities show that this is not straightforward. The proton affinity of a particular site change by 15–18 kcal/mol depending on whether it interacts with a QM or a MM residue. This means that to obtain a balanced description of different proton affinities all important residues still would have to be included in the QM part.

An important result from the modeling of the base protonation mechanism is that a certain configuration of the invariant network of charged amino acids results in a significant TS stabilization for cleavage of the C–C bond. If C6 is stabilized by other means than Lys93, this positive residue can keep its positive charge and form better interactions with the negative residues Asp91 and Asp96B. In total, when all effects of solvation have been included, this conformation leads to a barrier 8 kcal/mol lower than the model without enzyme residues.

For the remaining reaction mechanism, where the C–C bond is completely cleaved before the protonation occurs, only initial investigations have been done so far. This turns out to be the mechanism that appears most promising. A somewhat different procedure was followed for this mechanism. For the other mechanisms investigated the starting structures for the QM or QM/MM models were taken from the X-ray structure. For the investigation of this mechanism the starting structures were obtained from molecular dynamics treatments of the X-ray structure [38]. The best calculated barrier for this mechanism is obtained from QM/MM calculations, and is in the range 19–22 kcal/mol, which is in reasonable agreement with the experimental barrier of 15 kcal/mol, in particular considering the preliminary nature of the investigations. The origin of one part of the TS stabilization in this mechanism are the weakly bound water molecules that electrostatically interact with C6 at the TS. At the same time, Lys93 can keep the configuration that was shown for the base protonation mechanism to be advantageous. So far, it has not been possible to describe this mechanism using a pure QM model. As it turns out, in a QM modeling it is difficult to get the water molecules into their favorable C6 stabilizing positions in the transition state. One difference between QM and QM/MM models is the way weakly bound water molecules are treated. A truncated QM model runs the risk of introducing artificial hydrogen bonding positions that are already occupied in the enzyme. This problem can be treated by constraining the movement of the residues, as is done in Model 21, but at the risk of including other artificial effects if the restrictions are too severe.

All calculations in this study are made by direct optimization from X-ray structures or other protein structures provided. This means that stationary points are only detected if they are sufficiently close to the initial guesses. It also means that dynamic effects are neglected. Treatments including dynamics have been applied to ODCase [7, 38, 52] but the results are not consistent when it comes to the origin of the catalytic effect. The effect of dynamics on the present models cannot be easily estimated. However, it seems like a molecular dynamics treatment might be required to get the right structures for the stepwise mechanism with cleavage of the C–C bond without protonation.

The elucidation of the catalytic reaction mechanism of ODCase has proven to be the most difficult mechanistic problem so far encountered, even compared to a large number of metalloenzymes. In metalloenzymes the origin of the catalytic power is to a large extent the metal site, and the structure of the active site is more stable in many metalloenzymes than in ODCase. In ODCase there is no cofactor, so it is not obvious from which part of the active site the catalytic effect comes. The most likely mediator of the catalytic effect is the network of charged residues, also involving some water molecules. This might also be the source of the difficulties in describing the mechanism of ODCase, since the interactions between these groups are strong and the energy of different conformations must be reproduced accurately.

Acknowledgments We would like to thank Arieh Warshel and Marek Ŝtrajbl for kindly providing their structures of the ODCase reactants and intermediates. We would also like to thank the National Supercomputer Center (Sweden) for a generous grant of computer time at the SGI3800.

References

1. Radzicka A, Wolfenden R (1995) Science 267:90–92
2. Shostak K, Jones ME (1992) Biochemistry 31:12155–12161
3. Cui W, DeWitt JG, Miller SM, Wu W (1999) Biochem Biophys Res Comm 259:133–135
4. Miller BG, Smiley JA, Short SA, Wolfenden R (1999) J Biol Chem 274:23841–23843
5. Beak P, Siegel B (1976) J Amer Chem Soc 98:3601–3605
6. Lee JK, Houk KN (1997) Science 276:942–945
7. Lee T-S, Chong LT, Chodera JD, Kollman PA (2001) J Amer Chem Soc 123:12837–12848
8. Becke AD (1993) J Chem Phys 98:1372–1377
9. Becke AD (1993) J Chem Phys 98:5648–5652
10. Traut TW, Temple BRS (2000) J Biol Chem 275:28675–28681
11. Appleby TC, Kinsland C, Begley TP, Ealick SE (2000) Proc Natl Acad Sci USA 97:2005–2010

12. Miller BG, Hassell AM, Wolfenden R, Milburn MV, Short SA (2000) Proc Natl Acad Sci USA 97:2011–2016
13. Wu N, Mo Y, Gao J, Pai EF (2000) Proc Natl Acad Sci USA 97:2017–2022
14. Harris P, Poulsen J-CN, Jensen KF, Larsen S (2000) Biochemistry 39:4217–4224
15. Miller BG, Snider MJ, Wolfenden R, Short SA (2001) J Biol Chem 276:15174–15176
16. Miller BG, Butterfoss GL, Short SA, Wolfenden R (2001) Biochemistry 40:6227–6232
17. Porter DJT, Short SA (2000) Biochemistry 39:11788–11800
18. Siegbahn PEM (2003) Quart Rev Biophys 36:91–145
19. Siegbahn PEM, Blomberg MRA (2000) Chem Rev 100:421–437
20. Siegbahn PEM, Blomberg MRA (2001) J Phys Chem 105:9375–93864
21. Lundberg M, Blomberg MRA, Siegbahn PEM (2002) J Mol Model 8:119–130
22. Curtiss LA, Raghavachari K, Redfern RC, Pople JA (2000) J Chem Phys 112:7374–7383
23. Froese RDJ, Humbel S, Svensson M, Morokuma K (1997) J Phys Chem A 101:227–233
24. Siegbahn PEM (2001) Theor Chem Acc 105:197–206
25. Dunning TH Jr, Hay PJ (1976) Gaussian basis sets for molecular calculations. In: Schaefer HF (ed) Methods of electronic structure theory: Modern theoretical chemistry, vol 3. Plenum, New York, pp 1–28
26. Siegbahn PEM (1996) Electronic structure calculations for molecules containing transition metals. In: Prigogine I, Rice SA (eds) New methods in computational quantum mechanics: Advances in chemical physics, vol XCIII. Wiley, New York, pp 333–387
27. Schrödinger Inc. (1991–2000) Jaguar 4.1. Schrödinger, Portland, OR
28. Frisch MJ, Trucks GW, Schlegel HB, Scuseria GE, Robb MA, Cheeseman JR, Zakrzewski VG, Montgomery JA, Stratman RE, Burant JC, Dapprich S, Millam JM, Daniels AD, Kudin KN, Strain MC, Farkas O, Tomasi J, Barone V, Cossi M, Cammi R, Mennucci B, Pomelli C, Adamo C, Clifford S, Ochterski J, Petersson GA, Ayala PY, Cui Q, Morokuma K, Malick DK, Rabuck AD, Raghavachari K, Foresman JB, Cioslowski J, Ortiz JV, Baboul AG, Stefanov BB, Liu C, Liashenko A, Piskorz P, Komaromi, I, Gomperts R, Martin RL, Fox DJ, Keith T, Al-Laham MA, Peng CY, Nanayakkara A, Gonzalez C, Challacombe M, Gill PMW, Johnson BG, Chen W, Wong MW, Andres JL, Gonzales C, Head-Gordon M, Replogle ES, Pople JA (1998) Gaussian 98. Gaussian, Pittsburgh PA
29. Wiberg KB, Rablen PR, Rush DJ, Keith TA (1995) J Am Chem Soc 117:4261–4270
30. Wiberg KB, Keith TA, Frisch MJ, Murcko M (1995) J Phys Chem 99:9072–9079
31. Klamt A, Schuurmann G (1993) J Chem Soc Perk T 25:799–805
32. Barone V, Cossi M (1998) J Phys Chem A 102:1995–2001
33. Tannor DJ, Marten B, Murphy R, Friesner RA, Sitkoff D, Nicholls A, Ringnalda M, Goddard III WA, Honig B (1994) J Am Chem Soc 116:11875–11882
34. Schrödinger Inc. (2002) Qsite, FirstDiscovery Suite v2.0. Schrodinger, Portland, OR
35. Murphy RB, Philipp DM, Friesner RA (2000) Chem Phys Lett 321:113–120
36. Murphy RB, Philipp DM, Friesner RA (2000) J Comp Chem 21:1442–1457
37. Jorgensen WL, Maxwell DS, Tirado-Rives J (1996) J Am Chem Soc 118:11225–11236
38. Warshel A, Ŝtrajbl M, Villà J, Florián J (2000) Biochemistry 39:14728–14738
39. Singleton DA, Merrigan SR, Kim BJ, Beak P, Phillips LM, Lee JK (2000) J Am Chem Soc 122:3296–3300
40. Smiley, JA, Paneth P, O'Leary MH, Bell JB, Jones ME (1991) Biochemistry 30:6216–6223
41. Rishavy MA, Cleland WW (2000) Biochemistry 39:4569–4575
42. Phillips LM, Lee JK (2001) J Am Chem Soc 123:12067–12073
43. Smiley JA, Saleh L (1999) Bioorg Chem 27:297–306

44. Smiley JA, Hay KM, Levison BS (2001) Bioorg Chem 29:96–106
45. Miller BG, Snider MJ, Short SA, Wolfenden R (2000) Biochemistry 39:8113–8118
46. Miller BG, Traut TW, Wolfenden R (1998) Bioorg Chem 26:283–288
47. Warshel A, Florián J (1998) Proc Natl Acad Sci USA 95:5950–5955
48. Warshel A (1998) J Biol Chem 273:27035–27038
49. Villà J, Warshel A (2001) J Phys Chem B 105:7887–7907
50. Houk KN, Lee JK, Tantillo DJ, Bahmanyar S, Hietbrink BN (2001) Chem Bio Chem 2:113–118
51. Miller BG, Wolfenden R (2002) Annu Rev Biochem 71:847–885
52. Hur S, Bruice TC (2002) Proc Natl Acad Sci USA 99:9668–9673

Top Curr Chem (2004) 238:113–136
DOI 10.1007/b94541

Catalysis by Enzyme Conformational Change

Jiali Gao[1] (✉) · Kyoungrim Lee Byun[2] · Ronald Kluger[3]

[1] Department of Chemistry, Supercomputing Institute, and Digital Technology Center, University of Minnesota, Minneapolis, MN 55455, USA
gao@chem.umn.edu
[2] Computational Science, Korea Institute for Advanced Study, 130-722, Seoul, Korea
[3] Department of Chemistry, University of Toronto, Toronto, Ontario, M5S 3H6, Canada

Abstract An energy decomposition scheme is presented to elucidate the importance of the change of protein conformation substates to the reduction of activation barrier in an enzyme-catalyzed reaction. The analysis is illustrated by the reaction of orotidine 5′-monophosphate decarboxylase (ODCase), in which the catalyzed reaction is at least 10^{17} faster than the spontaneous reaction. Analysis reveals that the enzyme conformation is more distorted in the reactant state than in the transition state. The energy released from conformational relaxation of the protein is the main source of the rate enhancement. The proposed mechanism is consistent with results from site-directed mutagenesis where mutations remote from the reaction center affect k_{cat} but not K_M.

Abbreviations

AMI	Austin Model 1
EVB	Empirical valence bond
FEP	Free energy perturbation
MM	Molecular mechanics
ODCase	Orotidine 5′-monophosphate decarboxylase
OMP	Orotidine 5′-monophosphate
PMF	Potential of mean force
QM	Quantum mechanics
QM/MM	Combined quantum mechanical and molecular mechanical
TS	Transition state

1 Introduction

The ability of enzymes to accelerate reactions has been quantified by comparison with the rate of the corresponding uncatalyzed process. The example provided by orotidine 5′-monophosphate (OMP) decarboxylase (ODCase) is especially dramatic. In a classic study, Radzicka and Wolfenden showed that the rate enhancement (k_{cat}/k_{aq}) of the decarboxylation of OMP by ODCase is more than 10^{17} times compared to the decarboxylation of the model provided by *N*-methyl orotate (NMO) in water [1]. Although even greater rate enhancements have been observed in other enzyme systems [2], ODCase distinguishes itself from other enzymes by the fact that there is no direct participation of cofactors or metal ions, nor formation of a covalent intermediate with a group on the protein. Numerous experimental and computational studies have been carried out, aimed at understanding the catalytic process of ODCase. Yet, much is still to be learned about the catalytic power of ODCase. In this article, we present a view of ODCase catalysis on the basis of molecular dynamics free energy simulations. This model, which follows the idea of the Circe effect proposed by Jencks [3], is illustrated by an analysis of energy contributions due to the dynamical change of internal conformations of the enzyme along the reaction pathway in the active site of ODCase.

More than fifty years ago, in the absence of structural information, Pauling proposed that enzymes catalyze chemical reactions by specifically binding the transition state more strongly than binding the reactant state, consequently lowering the free energy of activation [4]. This remains the main concept in our understanding of enzyme catalysis today. With the development of modern transition state theory, the advent of computer simulation methods, and the availability of three-dimensional structures of enzymes and enzyme–substrate complexes, detailed energy calculations can now be carried out to provide an analysis of the specific means by which the activation barrier is lowered. Although much discussion has focused on transition state (TS) stabilization [5], dynamic fluctuations and internal motions of the enzyme are integral to its function and activity [6] and their contribution to catalysis remains to be fully appreciated.

Numerous experimental and theoretical studies have documented the direct correlation between internal motions of an enzyme and its activity. For example, the catalytic activities of many enzymes are closely associated with loop motions, which open and close the active site and also position key residues into contact with the substrate [7, 8]. X-ray crystal structures of ligand-bound and free enzymes show that substantial conformational changes can be induced by ligand binding and by chemical transformation during an enzymatic reaction [9, 10]. The reaction catalyzed by thymidylate synthase has been shown to couple directly with the protein's conformational changes along the multi-step reaction pathway [11, 12]. There is also a remarkable non-additive effect in enzyme activity from double mutations [13]. Such a non-additive effect on rate also occurs as a result of altering the structure of the substrate in ODCase [14]. In this article, we describe a computational approach to determine protein conformational energetic contributions to catalysis.

The comparison of non-enzymatic and enzymatic reactions can be calibrated in many ways. The calibration of ODCase provides a unique insight and sets the stage for remarkable progress in understanding the role of the protein. OMP is highly resistant to decarboxylation. In this case, no cofactor or any other small molecule (or protein side chain) has an ordinary catalytic role in the primary acceleration provided by the enzyme. Attempts to find ways in which the enzyme would use the addition of a nucleophile, or a conventional acid or base, led to mechanisms that were subsequently disproved. From the currently available X-ray structures of this enzyme it is clear that the catalysis is provided by the protein's ability to provide a specific environment that factors binding energy into components that drive the substrate toward the transition state. This form of catalysis takes advantage of the complexity, stability, and flexibility of the tertiary structure of the protein—the fundamental forces that were the subject of the original questions as to whether a pure protein could be a catalyst, which was settled in 1926 by Sumner's crystallization of urease. The methods of modern theoretical chemistry provide a unique opportunity to test the revelations of the crystal structure and the role of the protein.

There have been extensive studies of the ODCase reaction. Crystal structures of the enzyme complexes with several inhibitors from four different species show a remarkable similarity in the active site [15, 16, 17, 18], which consists of a network of charged residues (Lys42-Asp70-Lys72-Asp75b; we use the *thermoautotropicum* ODCase numbering throughout this paper) that are strictly conserved over 80 species [19]. Site-directed mutagenesis experiments by Wolfenden and coworkers showed that replacement of any of these charged residues essentially abolishes the enzyme's activity [20]. Moreover, mutations of residues that are far from the reactive site also significantly reduce the catalytic activity [14]. Several crystal structures of mutant ODCases have also been determined and they are not very different from the wild-type enzyme in the sense of small root-mean-square deviations [21, 22]. Mechanisms that involve formation of a zwitterion by protonation at the C2 carbonyl group [23], a carbene intermediate as a result of protonation at the C4 carbonyl group [24], and protonation or nucleophilic addition to C5 of the pyrimidine ring [25, 26], are unlikely because of a lack of acidic residues that could serve as proton donors or nucleophiles near C5 [27]. On the other hand, the hydrogen-bonding network in the active site appears to focus a negative charge in the vicinity of the substrate's carboxylate group, a stressful situation that could promote the required departure as a consequence of the Circe effect. Molecular dynamics simulations based on the structure of the *thermoautotropicum* ODCase, using a combined quantum mechanical and molecular mechanical (QM/MM) potential [16], and, in another work, using the empirical valence bond model [28], reproduced the experimental free energies of activation for the decarboxylation of OMP (or NMO) in the enzyme and in water. However, interpretation of these results has led to different conclusions on some aspects.

In order to focus on the factors that contribute catalysis, we present a free energy decomposition analysis and define energy terms. The theory is then applied to the reaction of ODCase to arrive at the origins of the observed catalysis in this remarkable enzyme. The article concludes with a summary of major findings.

2 Free Energy Decomposition

We begin by considering the following thermodynamic cycle, which relates the experimentally determined free energies of activation for the uncatalyzed reaction in water ($\Delta G^{\neq}_{aq}$) and the catalyzed reaction in enzyme ($\Delta G^{\neq}_{cat}$), the binding free energy of the substrate to form the Michaelis complex ($\Delta G_b(E'S)$), and the apparent binding free energy for the transition state ($\Delta G_b(E''S^{\neq})$).

$$\begin{array}{ccccc} & E+S & \xrightarrow{\Delta G^{\ddagger}_{aq}} & E+S^{\ddagger} & \\ \Delta G_b(E'S) & \downarrow & & \downarrow & \Delta G_b(E''S^{\ddagger}) \\ & E'S & \xrightarrow{\Delta G^{\ddagger}_{cat}} & E''S^{\ddagger} & \end{array} \qquad \text{(a)}$$

$$\Delta\Delta G^{\neq} = \Delta G^{\neq}_{cat} - \Delta G^{\neq}_{aq} = \Delta G_b\left(E''S^{\neq}\right) - \Delta G_b(E'S) \qquad (1)$$

In this scheme, E and S represent the enzyme and substrate, respectively, whereas E' and E'' indicate different conformational substates of the enzyme induced by "binding" of the substrate and the distorted structure of the substrate in the transition state. In the Michaelis-Menten mechanism, the binding free energy of the substrate $\Delta G_b(E'S)$ can be related to K_M by $\Delta G_b(E'S)=-RT\ln(1/K_M)$. It is important to note that $\Delta G_b(E''S^{\neq})$ in Eq. 1 is an *apparent* binding free energy of the species corresponding to the transition state (TS) of the enzymatic reaction, which is not directly measurable but is always inferred from $(k_{cat}/K_M)/k_{aq}$. Although it is often convenient to use the term of transition state binding, the activated complex $E''S^{\neq}$ created along the enzymatic reaction pathway may differ from the actual species produced by binding of $S^{\neq}$ from solution. In the latter case, the protein conformation and the presence of water molecules in the active site may be markedly different than the former, as seen in X-ray crystal structures of enzymes to which substrates and transition state analogs are bound [29, 30]. Therefore, the reduction of the activation barrier in the enzymatic pathway cannot be simply rationalized by the difference in the "binding" of the TS and the reactant substrate.

In other words, $\Delta\Delta G^{\neq}$ in Eq. 1 is the energetic definition of enzyme catalysis, which represents the difference in free energy of activation between the catalyzed reaction and an equivalent, uncatalyzed model reaction, which is related to the ratio of the rate constants of the catalyzed and the uncatalyzed reactions (k_{cat}/k_{aq}). This definition has been used by many workers in numerous studies [31, 32, 33]. Often, the uncatalyzed reaction in aqueous solution is used as the reference state, although it is important to study the same reaction in the gas phase to gain an understanding of solvation effects and the intrinsic reactivity of the model reaction. Interestingly, since, by definition, $\Delta\Delta G^{\neq}\leq 0$ for the catalyzed reaction, Eq. 1 can always be interpreted in terms of TS stabilization [5]. Thus, modern computational approaches must go beyond the statement of transition state stabilization by elucidating the specific means by which such a stabilization is achieved [28, 34]. It is important to identify specific energy terms that contribute to the lowering of the free energy of activation in the enzymatic environment that give rise to a greater apparent binding affinity for the TS than the reactant substrate.

One can readily identify three main factors that contribute to the free energy change $\Delta\Delta G^{\neq}$ in Eq. 1. First is transition state stabilization, where the enzyme provides greater stabilizing interactions with the distorted structure of the substrate in the transition state than solvent water does. The second and third effects can both be thought of as ground state destabilization, in which the substrate and the enzyme, respectively, adopt more distorted conformational substates in the Michaelis complex than they do separately in water [10, 35]. An example of the substrate's conformational distortion is the Claisen rearrangement of chorismate, which assumes a diaxial conformation in the active site of chorismate mutase.This is about 5–6 kcal/mol less stable than the diequatorial conformation favored in solution [36, 37, 38]. The distortion of the enzyme itself, induced by binding of the substrate, whose energy is subsequently released in the enzymatic transition state, is illustrated here for the decarboxylation reaction of ODCase.

To distinguish contributions between transition state stabilization and enzyme conformational change, these energy terms must be defined and computed (see below), which requires the separation of the overall barrier reduction $\Delta\Delta G^{\neq}$ into specific components. To achieve this goal, we partition the binding free energy of the substrate $\Delta G_b(E'S)$ into two terms, as shown in Fig. 1. The first term is protein distortion energy, $\Delta G_P(E')$, resulting from conformational changes in going from the unbound enzyme, E, to that in the Michaelis complex, E′, that is induced by substrate binding. Since the

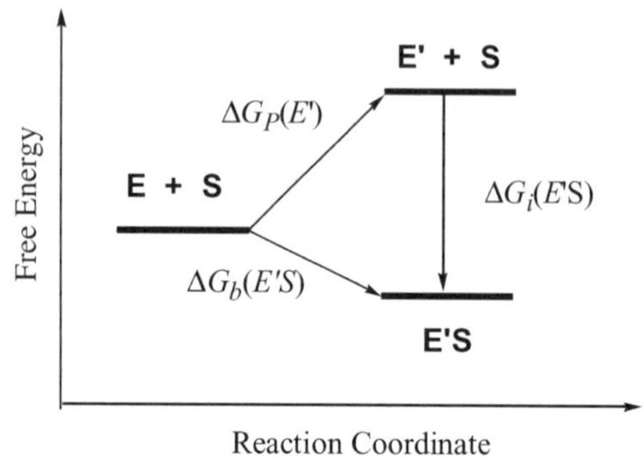

Fig. 1 Schematic representation of binding free energy decomposition used in the analysis discussion in this paper

unbound free enzyme in solution is necessarily a more stable state, the term $\Delta G_P(E')$ must be positive. The interaction between the substrate and the distorted enzyme is defined as the intrinsic binding free energy $\Delta G_i(E'S)$. Therefore, the observed free energy of binding is the sum of the two contributions (Fig. 1):

$$\Delta G_b(E'S) = \Delta G_P(E') + \Delta G_i(E'S) \tag{2}$$

A similar decomposition can be made for the apparent binding free energy of the transition state species:

$$\Delta G_b\left(E''S^{\neq}\right) = \Delta G_P(E'') + \Delta G_i\left(E''S^{\neq}\right) \tag{3}$$

where the notation E'' specifies the enzyme conformation substate at the TS of the substrate–protein complex. Consequently, the change in free energy of activation (Eq. 1) in the enzyme originates from two important factors:

$$\Delta\Delta G^{\neq} = \Delta\Delta G^{\neq}_{ES} + \Delta\Delta G^{\neq}_{PP} \tag{4}$$

where $\Delta\Delta G^{\neq}_{ES}$ and $\Delta\Delta G^{\neq}_{PP}$ are, respectively, the difference in intrinsic binding free energy between the TS and the reactant state, and the difference in protein distortion energy. They are defined below:

$$\Delta\Delta G^{\neq}_{ES} = \Delta G_i\left(E''S^{\neq}\right) - \Delta G_i(E'S) \tag{5}$$

$$\Delta\Delta G^{\neq}_{PP} = \Delta G_P(E'') - \Delta G_P(E') \tag{6}$$

The term $\Delta\Delta G^{\neq}_{ES}$ in Eq. 5 is concerned with the differential effects of the solvent and the enzyme environment on the transition state and reactant state. If the enzyme provides stronger stabilizing interactions to the transition state than the reactant state relative to the difference in water, such an effect is defined as TS stabilization, and it corresponds to a negative value in$\Delta\Delta G^{\neq}_{ES}$. Although, in principle, Eq. 6 can include another term that specifies the energy difference due to the change in substrate conformation upon binding, this term is not significant in the ODCase reaction, and, thus, for clarity in the present discussion, we only list in Eq. 6 the conformational energy change of the protein in going from the Michaelis complex to the transition state. The value of $\Delta\Delta G^{\neq}_{PP}$ can be either positive or negative depending on the relative distortion of the enzyme when the substrate is in the reactant state or at the transition state. Since this free energy term is not directly associated with protein–substrate interactions, it cannot be classified as "transition state stabilization", although Eq. 6 still makes contributions to the reduction of the activation barrier.

Furthermore, because both terms on the right-hand side of Eq. 6 are positive quantities by definition, their contributions to enzyme catalysis are defined as ground state destabilization. If the distortion of the enzyme is greater in the Michaelis complex than that in the transition state, relaxation of the enzyme conformation along the reaction coordinate leads to lowering in the overall activation energy. Obviously, the enzyme could have an even more distorted conformation at the transition state, induced by greater interactions between the enzyme and the transition state. In this case, the $\Delta\Delta G^{\neq}_{PP}$ term makes "anti-catalytic" contribution and is positive.

It should be noted that the term of ground state destabilization does not simply refer to the situation where the total free energy of the enzyme–substrate complex is higher than that of the apo-enzyme and free substrate in solution. Ground state destabilization describes the case that part of the system, either the substrate or the protein, is more distorted, and hence has higher energy than the free, unbound state. The overall free energy of the Michaelis complex is still lower than the unbound state (note that the binding constant is dependent on the choice of the standard state). In some discussions [28, 34, 39], this difference was not clearly distinguished and the specific energy terms were not computed, leading to the incomplete impression that ground state destabilization must raise the overall energy of the Michaelis complex without changing the energy of the transition state (see Fig. 6 of [39]) . An analysis of the energy component is necessary to understand specific contributions to the overall reduction of the activation free energy.

At this point, we return to the discussion of the significance of enzyme conformational change in catalysis, which has long been recognized [40]. Jencks proposed that through the Circe effect [3] part of the intrinsic binding energy can be used to compensate for destabilizing interactions in the "reactive part" of the substrate that is enforced by binding of the entire substrate. The analysis we present is an extension of this proposal in that a fraction of the intrinsic binding energy is stored in the form of conformational distortion energy of the protein in the Michaelis complex. This energy is then released at the transition state of the enzyme–substrate complex, resulting in the reduction of the activation barrier. This mechanism is different than the induced fit proposal, which is concerned with the change of enzyme conformation to better fit and bind the substrate so that protein–substrate interactions are enhanced [40].

3 Computational Approach

It would be very difficult to determine experimentally the free energy terms of Eqs. 5 and 6, but this can be done by free energy calculations. To ensure computational accuracy, both the potential energy surface and the simulation method must be validated.

3.1 Potential Energy Function

A prerequisite for computational study of enzymatic reactions is an accurate description of the potential energy surface that involves the breaking and forming of chemical bonds. In principle, the most accurate approach is quantum mechanics, and a computationally feasible model is the combination of quantum mechanics (QM) with molecular mechanics (MM) [41, 42], in which the active site is represented by QM and the remainder of the protein/solvent system is approximated by MM. Thus, the effective Hamiltonian for a hybrid QM and MM system is given as follows:

$$\hat{H}_{eff} = \hat{H}_{qm} + \hat{H}_{qm/mm} + \hat{H}_{mm} \tag{7}$$

where $\hat{H}_{qm}$ is the Hamiltonian for an isolated QM region in the gas phase, $\hat{H}_{qm/mm}$ is the interaction Hamiltonian between the QM and MM regions, and $\hat{H}_{mm}$ molecular mechanics energy of the MM region. Both $\hat{H}_{qm}$ and $\hat{H}_{qm/mm}$ contain electronic degrees of freedom, whereas $\hat{H}_{mm}$ does not include electrons explicitly. The total potential energy of the system is given in Eq. 8:

$$E_{pot} = <\Psi\left|\hat{H}_{eff}\right|\Psi> \tag{8}$$

where ψ is the wave-function of the system.

Although the idea of combining quantum mechanics with force fields to study enzyme reactions has been described as early as 1976 by Warshel and Levitt, the use of combined QM/MM potentials in molecular dynamics and Monte Carlo simulations was not carried out until about the 1990s [42, 43, 44], largely because its validity was not thoroughly tested up to that time [42, 45, 46]. Warshel and coworkers later developed a useful, empirical valence bond (EVB) model [31], employing MM force fields to increase computational speed and to make it convenient for parameter calibration. Nevertheless, combined QM/MM methods have the advantage of computational accuracy by treating the reactive part of the system explicitly with quantum mechanics, and computational efficiency by approximating the much larger part of the system with a force field. Furthermore, the electronic polarization of the active site is naturally included in QM calculations. In fact, combined QM/MM methods have emerged as the best approach for studying enzymatic reactions [47].

In the empirical valence bond model [31], the potential energy surface is typically non-linearly related to two model states, called effective *diabatic* states corresponding to the reactant and the product bonding characters, Eq. 9. Although additional diabatic states, important for describing the transition state, can be included and have often been discussed, the computational complexity makes it difficult in practical use for the study of enzyme reactions.

$$E_{EVB}=\frac{1}{2}\left\{(E_R+E_P)-\sqrt{(E_R-E_P)^2+4\varepsilon_{12}^2}\right\} \tag{9}$$

In Eq. 9, E_R and E_P are energies for the reactant and product states, which are represented by Morse potentials for the dissociating bonds plus MM terms for the rest of system, and ϵ_{12} is the electronic coupling, which is typically treated as a constant or a simple exponential function of some geometrical variables. As can be seen, an important feature of the EVB model is the mixture of the diabatic states, which is analogous to the valence bond treatment of the hydrogen molecule using the expression of Eq. 9, a flavor of quantum mechanics. Nevertheless, electronic structure is not explicitly treated (i.e., no wave-functions) in contrast to combined QM/MM potentials. Although empirical, the EVB method strives for accuracy by fitting three key parameters in the Hamiltonian (two diagonal and one off-diagonal matrix elements) to three experimental observables, corresponding to the dissociation energies of the reactant and product states and the free energy of activation. These parameters are often calibrated to reproduce experimental activation energy for the model reaction in water, and then, they are used in enzyme calculations without further adjustments [31].

Before we proceed, it is important to clarify the procedure of parameter fitting, and its validation. The quantity to be compared in analyzing the origin of enzyme catalysis is the change in free energy of activation in the enzyme relative to an equivalent, uncatalyzed reaction in water. Therefore, it is reasonable to assume that the computational model that is parameterized for the aqueous reaction and can reproduce the reduction of free energy barrier in the enzyme can yield insights on enzyme catalysis. Nevertheless, we emphasize that it is equally critical to validate the computational model for the model reaction in the gas phase to shed light on the intrinsic properties on structure and energy along the entire reaction coordinate by comparison with gas phase experimental and high-level ab initio results. Thus, only after the performance of the computational model is validated on gas-phase properties, can one proceed to examine condensed phase effects on the reaction barrier, in particular, the aqueous solvation effect. The final step is to model the enzymatic process using exactly the same model that is used to describe the gas phase and aqueous reactions.

The semi-empirical AM1 model that we use has been validated for several decarboxylation reactions [16, 48, 49, 50] and for solute–solvent and ligand–protein interactions by considering bimolecular complexes and free energies of solvation [42, 46, 49]. It is also important to point out that combined QM/MM models, even at the semi-empirical level, provide a good description of the effects of solvation and protein environment on the substrate's reactivity because charge polarization by the environment is explicitly treated by QM calculations [51, 52, 53]. This is in contrast to empirical functions in which atomic partial charges are typically fixed throughout the calculation without responding to the dynamic fluctuations of the enzyme and solvent of the condensed phase system (of course, fixed charge approximation is not a limitation of the EVB model, but a polarizable EVB has rarely been applied to enzyme reactions) [28].

3.2 Potential of Mean Force

The starting point to understand enzyme catalysis using computational methods is to determine the potentials of mean force (PMF) along a predefined reaction coordinate for both the catalyzed and uncatalyzed reactions in water. The PMF yields the free energy of activation, which provides validation to the potential energy function used to describe the chemical process and the interactions with the environment (solvent and/or enzyme). For the OMP decarboxylation by ODCase, the reaction coordinate is conveniently defined as the distance between the C6 atom of the pyrimidine moiety and the carbon atom of the carboxylate group of OMP substrate [16]. Although there are two main approaches in free energy simulations, namely the free energy perturbation method and the umbrella sampling technique [54], we employ the latter approach in our studies of enzymatic reactions to obtain the PMF, which is related to the probability density $\rho(R_c)$ at a given value of the reaction coordinate, R_c:

$$W(R_c) = -RT \ln \rho(R_c) + C \tag{10}$$

where C is an arbitrary normalization constant, R is the gas constant, and T is temperature. In practice, the umbrella sampling calculation is carried out by modifying the potential energy surface with the addition of a biasing potential, which ideally is the negative of the PMF, $W(R_c)$ [55]. Therefore, protein conformations at high energy regions including the transition state can be adequately sampled in molecular dynamics simulations. It is important to note that $W(R_c)$ is the free energy of the system that includes both protein internal energy and protein–substrate interaction energy.

3.3 Free Energy Decomposition Analysis

The intrinsic binding free energy of the substrate corresponds to the free energy of transfer from water into the enzyme active site, which is related to the free energies of "solvation" of the substrate in these two environments. Here, we use the generic term "solvation" to describe the free energy of transfer from the gas phase into a specific environment. Computationally, the free energy perturbation (FEP) method has been established as an ideal approach for these calculations [54]:

$$\Delta G_{sol}^{0\to 1} = -RT\ln < e^{-(V_1-V_0)/RT} >_{V0} \tag{11}$$

where $\Delta G_{sol}^{0\to 1}$ is the free energy difference between state 0 and 1, V_0 and V_1 are potential energies of the two states, and the brackets $<...>_{V0}$ indicates the ensemble average over the potential surface of state 0. Typically, the calculation of solvation free energy is divided into two steps, first by electrostatic charge annihilation, followed by a "mutation" of the van der Waals spheres of atoms into nothing [54]. The latter term is often related to the free energy of cavitation, and it makes relatively small contributions to the total free energy of solvation for an ionic molecule. The difference in van der Waals contribution for the substrate in water and in the enzyme would be even smaller and is considered to have minor effects on the rate acceleration of ODCase catalysis.

4 Results

4.1 Potential of Mean Force

We have computed the potentials of mean force for the decarboxylation of *N*-methyl orotate (NMO) ion in water, which was used by Wolfenden as the reference reaction, and OMP substrate in ODCase using a combined QM/MM method in Monte Carlo and molecular dynamics simulations (Fig. 2) [16]. A separate calculation has been performed for the reaction of OMP in the gas phase, which yielded a reaction profile similar to that of NMO in the gas phase, indicating that there is no self-catalysis by the auxiliary phospho-

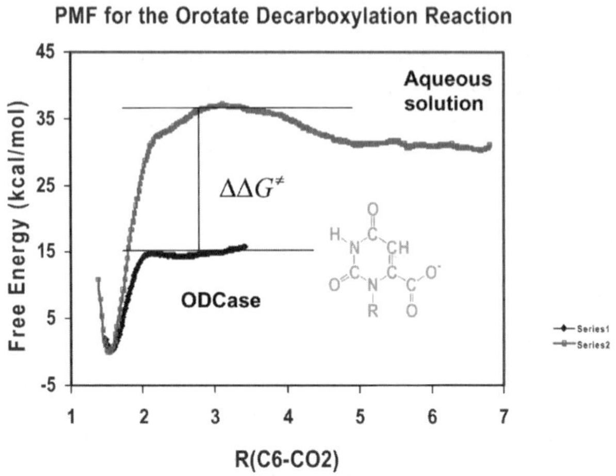

Fig. 2 Computed potential of mean force for the decarboxylation reaction of OMP in water and in the wild-type enzyme ODCase

ribosyl group, consistent with experiment [56]. The QM method used in these calculations is the semi-empirical Austin Model 1 (AM1) [57], which yields an energy of reaction of 35.5 kcal/mol in the gas phase, in excellent agreement with a value obtained at the MP2/6-31+G(d) level of theory (36.4 kcal/mol) [24]. In water, the calculated $\Delta G^{\neq}_{aq}$ is 37.2 kcal/mol, compared with the experiment value of 38.5 kcal/mol [1]. Interestingly, there is little solvent effect on the decarboxylation reaction, which has also been noticed in the work of Lee and Houk using a continuum salvation model in ab initio calculations [24].

The free energy of activation for the decarboxylation of OMP in ODCase was estimated to be 14.8 kcal/mol, a reduction of 22.4 kcal/mol relative to the uncatalyzed reaction in water, which is in agreement with experiment (15.2 kcal/mol from k_{cat}) [1]. This corresponds to a computed rate enhancement of k_{cat}/k_{aq}=2.6×10^{16}, which may be compared with the experimental value of 1.7×10^{17}. Thus, combined QM/MM simulations yield results in agreement with ab initio or experimental data for the decarboxylation reaction in the gas phase, in water, and in the enzyme ODCase, and thus, we can proceed to analyze the computational results to gain insights into the origin of catalysis [16].

Using the EVB approach, which was calibrated to the aqueous reaction [28], Warshel et al. also reproduced the experimental free energy of activation for the catalyzed (OMP) reaction in ODCase. The EVB energy profile and barrier for the decarboxylation reaction in the gas phase was not reported. Although the computed barrier heights for the enzyme reaction are similar in the two computational studies [16, 28], a major difference is that the

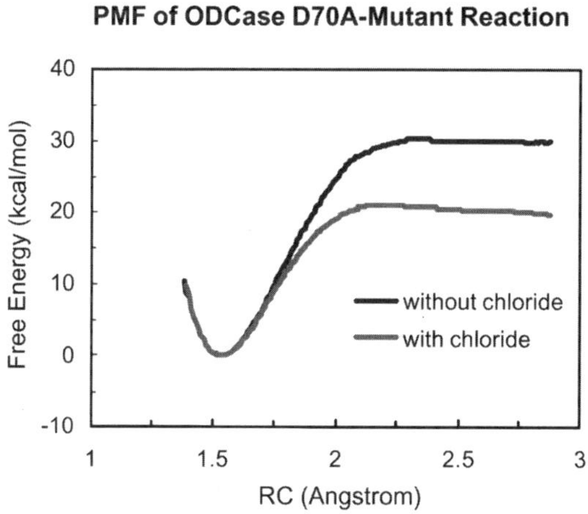

Fig. 3 Computed potentials of mean force for the decarboxylation reaction of OMP in the Asp70Ala mutant with and without a chloride ion in the active site

product, the C6-deprotonated UMP, is strongly stabilized by the enzyme to a free energy similar to that of the reactant state (OMP–ODCase complex) in the work of Warshel et al. [28], whereas this anion is not stable and is essentially the transition state of the decarboxylation reaction [16]. Early calculations show that the anion of the decarboxylation reaction is not a stable intermediate [16, 24]. Warshel [28] made an important contribution to the study of the ODCase reaction by analyzing the ionization states of various charged residues in and near the active site, using a model analogous to a continuum dielectric representation of the system and the theory of linear-response approximation (PDLD/s-LRA). In particular, it was found that the substrate OMP, Asp20, Glu25, Lys42, Asp70, Lys72, Asp75b, Glu78b, Lys82b, His98, and Arg203 are all ionized, confirming the same ionization state used in [16] on the bases of examination of interactions in the X-ray structure and available experiment information [16].

We have carried out two additional calculations: the first is the Asp70Ala mutant, and the second is the same mutant, but with the inclusion of a chloride ion in the active site (Fig. 3). The chloride ion was discovered in the Asp90Ala mutant crystal structure at the site where the Asp70 carboxylate would reside [21]. Further, it was found that incubation of OMP substrate in the crystal resulted in the conversion to UMP product during the X-ray diffraction study [21]. Based on these experimental findings, these two simulations were designed and the computed potentials of mean force are displayed in Fig. 3. The computed free energy of activation for the OMP decarboxylation in the Asp70Ala ODCase mutant is 30.2 kcal/mol, and the inclusion of a chloride ion in the active site of this mutant lowers the free energy barrier to 21.0 kcal/mol. Recall that the free energies of activation for the wild-type enzyme and the uncatalyzed reaction in water are computed to be 14.8 and 37.2 kcal/mol, respectively. Thus, we have a system that encompasses a gradual lowering of the free energy of activation by adjusting the electrostatic environment of the active site.

4.2 Intrinsic Binding Free Energy and Enzyme Conformational Energy

The electrostatic component of $\Delta\Delta G^{\neq}_{ES}$ was obtained by computing the electrostatic free energies of "solvation" for the substrate and its corresponding transition state species in water and in the enzyme through FEP simulations [16]. Table 1 summarizes the results, which reveal that the $\Delta\Delta G^{\neq}_{ES}$ term only lowers the free energy barrier by −2.2 kcal/mol for the enzymatic reaction compared to the uncatalyzed reaction in water, which is far from the total barrier reduction (−22.4 kcal/mol). We note that this does not contradict the concept of overall transition state stabilization in the catalyzed reaction relative to the uncatalyzed process because there is significant contribution from the $\Delta\Delta G^{\neq}_{PP}$ term. It has been suggested that $\Delta\Delta G^{\neq}_{ES}$ must reproduce the *entire* catalytic effect [28], but, in fact, it would be surprising if $\Delta\Delta G^{\neq}_{ES}$ is identical to $\Delta\Delta G^{\neq}$, which implies that the protein does not undergo conformational change, or has the same energies during the reaction. The small computed value of $\Delta\Delta G^{\neq}_{ES}$ indicates that the largest contribution to the reduction of the free energy barrier in ODCase is not due to enhanced electrostatic stabilization at the TS. Instead, it originates from the relaxation of the enzyme conformation from a highly distorted state in the Michaelis complex to a less distorted conformation in the transition state. Rearranging Eq. 4, we obtain an estimate of the contribution from enzyme conformation energy:

$$\Delta\Delta G^{\neq}_{PP} = \Delta\Delta G^{\neq} - \Delta\Delta G^{\neq}_{ES} = -22.4 - (-2.2) = -20.2\text{kcal/mol} \quad (12)$$

Table 1 Computed electrostatic free energies (kcal/mol) of transfer for the orotate group in water (w) and in the enzyme ODCase (p) from the gas phase (g). $\Delta\Delta G^{\neq}$ is obtained from potentials of mean force calculations (Fig. 2) and relative free energies are given with respect to the reactant state energy

	Reactant State	Transition State
$\Delta G^{g\to w}_{sol}$	−66.7	−57.5
$\Delta G^{g\to p}_{sol}$	−48.9	−41.9
$\Delta\Delta G^{w\to p}_{sol}$	17.8	15.6
$\Delta\Delta G^{\neq}_{ES}$	0.0	−2.2
$\Delta\Delta G^{\neq}$	0.0	−22.4
$\Delta\Delta G^{\neq}_{PP}$	0.0	−20.2

Note that the van der Waals interactions are not included in the $\Delta\Delta G^{\neq}_{ES}$ term; however, they are expected to be small in going from the reactant to the transition state in the ODCase reaction because of the relatively small changes in the substrate's structure, other than along the C–C distance of the reaction coordinate [28].

It is important to note that $\Delta\Delta G^{\neq}_{PP}$ is not computed directly. However, both $\Delta\Delta G^{\neq}$ and $\Delta\Delta G^{\neq}_{ES}$ are well-defined and can be computed accurately [54, 58]. Even if each of the computed $\Delta\Delta G^{\neq}$ and $\Delta\Delta G^{\neq}_{ES}$ terms has an error of 5 kcal/mol, which is very unlikely [54, 58], it is clear that changes in the protein's conformation from a more distorted reactant state to a less destabilized transition state will make major contributions to the reduction of the activation barrier. This is especially significant in understanding the ODCase mechanism.

5
The Mechanism of the ODCase Reaction

In analyzing the origin of enzyme catalysis, Warshel and others have advocated the importance of comparing the enzymatic reaction with a reference reaction in water [32]. In addition, it is also necessary to study the reference reaction in the gas phase in order to understand the intrinsic reactivity and the effect of solvation. Thus, to understand enzyme catalysis fully, we must compare results for the same reaction in the gas phase (intrinsic reactivity), in aqueous solution (solvation effects), and in the enzyme (catalysis). This is not possible when there is no model reaction for the uncatalyzed process in the gas phase and in water, or if the uncatalyzed reaction is a bimolecular process as opposed to a unimolecular reaction in the enzyme active site. None of these problems apply to the ODCase reaction. Furthermore, OMP decarboxylation is a unimolecular process, both in water and the enzyme, providing an excellent opportunity to compare *directly* the computed free energies of activation [1]; this is the approach that we have undertaken [16]. Warshel et al. used an ammonium ion–orotate ion pair fixed at distances of 2.8 or 3.5 Å as the reference reaction in water to mimic an active site lysine residue [32].

Can a ground state destabilization mechanism be possible in ODCase? To answer this question, we first examine the possibility of TS stabilization, which can be achieved in many enzymes by changing electrostatic forces and hydrogen bonding between the substrate and the enzyme along the reaction path [33, 59]. In the case of the OMP decarboxylation, the change of the electrostatic feature is a shift of the anionic charge from the C6-carboxylate group to a rather delocalized C6 carbenium ion of the pyrimidine ring. In order to stabilize the TS, a cationic residue must somehow specifically interact with the C6 carbenium ion rather than the carboxylate ion. Lys72 is a likely candidate, which in fact migrates by about 2 Å from a hydrogen-bonding position to the 2′-hydroxy group of the ribose to interact with the C6 anion. There are also other changes including Asp70 and Asp75b, to which Lys72 forms two other hydrogen bonds. If a strong electrostatic stabilizing effect is present, a very negative value in the $\Delta\Delta G^{\neq}_{ES}$ term would have been obtained. Yet, we obtained a value of only −2.2 kcal/mol.

We compared the structures obtained from molecular dynamics QM/MM simulations with that of the X-ray structure complexed with 6-aza-OMP transition state analog [16]. When doing calculations of enzymatic reactions, we took great care to make sure that the structures from the computational model are consistent with those reported in the crystal. The tests provided by such controls give us a high degree of confidence in the results and conclusions we are able to draw. In Fig. 4, the computed structures have been superimposed with the X-ray structure by matching the backbone atom positions of residues 30–200. The transition state configuration from the QM/MM simulation closely resembles that of the X-ray structure, whereas there is greater conformational distortion of the enzyme in the reactant state (Michaelis complex). Most significant are changes of the side-chain positions of residues Lys72, Asp70, and Asp75b among others, in a range of 1.5 to 2 Å. It is interesting to notice that the phosphoribosyl group is found essentially superimposed with the X-ray structure for both the reactant and transition state, although they are not included as part of the least square fit in making the structure superposition. In the reactant state (Fig. 4a), the orotate ring extrudes forward to minimize electrostatic repulsion with Asp70. Lys72 forms three hydrogen bonds with Asp72, Asp75b, and the 2′-hydroxy group (Fig. 5). The last hydrogen bond is replaced by interactions with the C6 carbanion in the TS as a result of 2 Å conformation change (Fig. 6). The distance between the side-chain N atom of Lys72 and the carbonyl group of the substrate is about 4.5 Å in the reactant state and 2.8 Å to the C6 anion in the TS. Similar changes have been found in molecular dynamics simulations by Hur et al. [27]

The change in protein conformation illustrated for the reaction has important implications in the reduction of the observed free energy barrier. In fact, it is a key to understanding the mechanism of the ODCase reaction. Figure 7 shows the free energy components indicated by arrows, depicting the protein distortion energy and intrinsic binding energy. The sum of the two contributions gives the experimental binding free energy (shown in blue). The decomposition analysis was done for both the reactant state and for the transition state for comparison. The computed free energies are given in Table 1. The formation of the Michaelis complex has an observed binding free energy of about 8 kcal/mol (from experimental K_M) [1], which has two contributing factors: (1) the intrinsic binding free energy, and (2) the protein conformational distortion energy. The intrinsic binding free en-

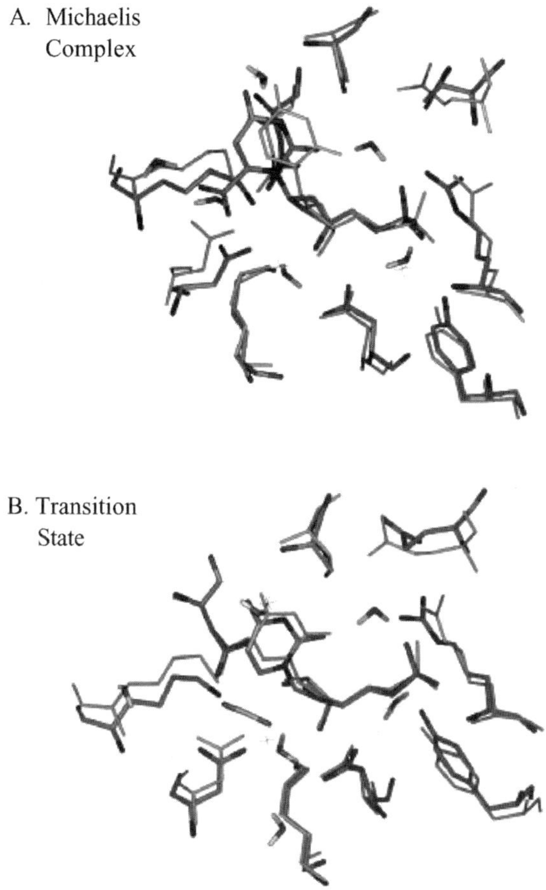

Fig. 4a,b Comparison of the structures of the Michaelis complex (**a**) and transition state complex (**b**) from molecular dynamics QM/MM simulations of Wu et al. with the X-ray structure (*gray*) of ODCase complexed with 6-aza-OMP [16]. Both simulation structures, which were a snapshot of the system after at least 200 ps molecular dynamics simulations, were superimposed with the X-ray structure on the basis of optimal fit of backbone atoms for residues from 30 through 200. **a** Seen in the reactant state are significant side-chain conformational change, distorted away from the X-ray structure, while the substrate also experience conformational stress. The distortion is widespread, away from and close to the reactive orotate group. Two water molecules were found to form hydrogen bonds with the carboxylate group of OMP in the simulation. **b** The simulated structure for the transition state is in remarkable agreement with the X-ray configuration both for the enzyme and the substrate, suggesting that the protein is less stressed. The two water molecules noted above diffuse away by more than 5 angstroms

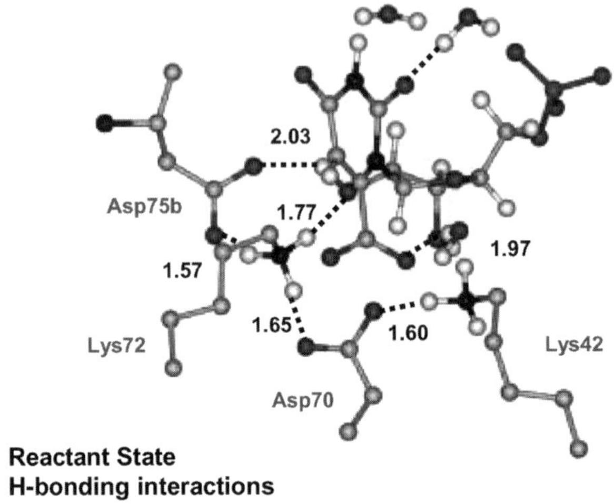

Fig. 5 Hydrogen bonding interactions for the reactant state. Distances are given in angstroms

ergy of the orotate group was obtained from free energy perturbation calculations for the reactant and transition state (Table 1). The protein distortion free energy cannot be computed directly, but its change can be derived from the computed free energies of activation using Eq. 4. At the TS, the intrinsic $E''S^{\neq}$ interaction is enhanced by −2.2 kcal/mol relative to the $E'–S$ state

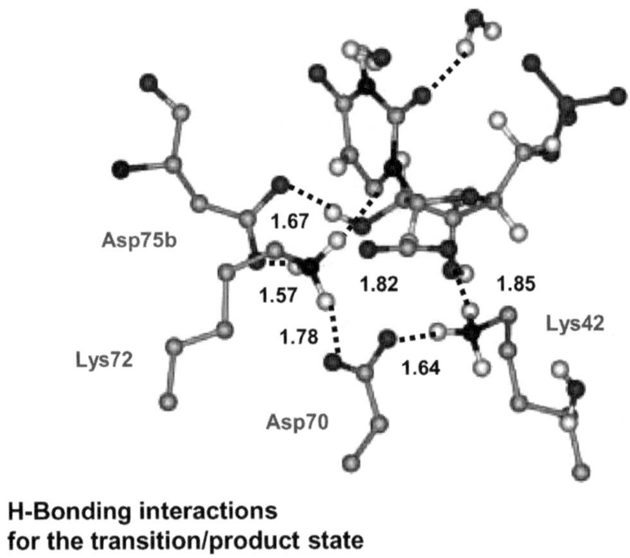

Fig. 6 Hydrogen bonding interactions for the transition state. Distances are given in angstroms

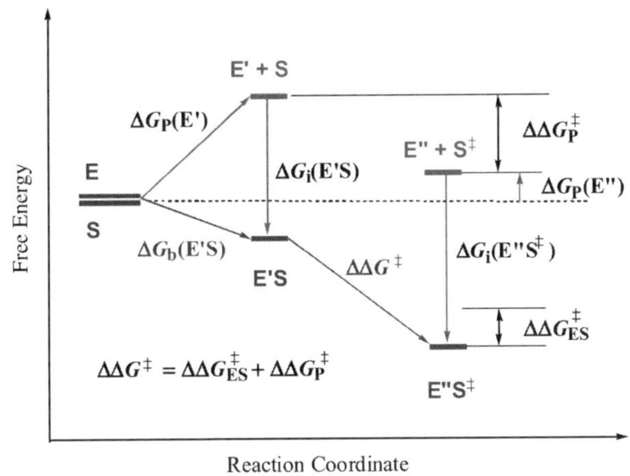

Fig. 7 Schematic illustration of the origin of catalysis by the enzyme ODCase. *Arrows* indicate the free energy decomposition of Eqs. 2 and 3 that separate the binding free energy into a protein distortion term and an intrinsic substrate–protein interaction component. The figure shows that the protein has a smaller distortion energy than that in the reactant state, so the change in protein conformational energy provides the predominant contribution to the lowering of the activation free energy of the catalyzed reaction

(Table 1). At the same time, the protein becomes more relaxed with a much smaller distortion free energy (about 20 kcal/mol) than that in the E' state. Figure 7 also illustrates a mechanism in which the reactant state of the protein is more destabilized in the Michaelis complex than the transition state. The destabilization energy at the reactant state is subsequently released during the chemical step through protein conformation change to give the remarkable apparent binding affinity for the transition state.

We attribute the large protein distortion in the reactant state to local electrostatic repulsions of the carboxylate group of orotate and Asp70, and the hydrophobic environment of the binding pocket surrounding the orotate moiety (Fig. 8). Clearly, this is not the sole source of electrostatic effects. This, however, should not be confused by the argument that such repulsive interactions would have led to protonation of one of the carboxylate groups at equilibrium [28]. This statement would have been correct if there were no other interactions in the active site. What such an argument overlooks is that there is a delicate balance of electrostatic interactions that also provides stabilization of the anions by a network of ionic interactions in the active site (Lys42-Asp70-Lys72-Asp75b). Specifically, the two lysine residues (Lys42 and Lys72) provide stabilizing contributions to compensate repulsive interactions. The repulsion from the carboxylate groups pushes Asp70 along with Lys72 away from the equilibrium positions in the X-ray structure (with the 6-aza-UMP transition state analog) by about 1.5 to 2 Å (Figs. 4–6). This re-

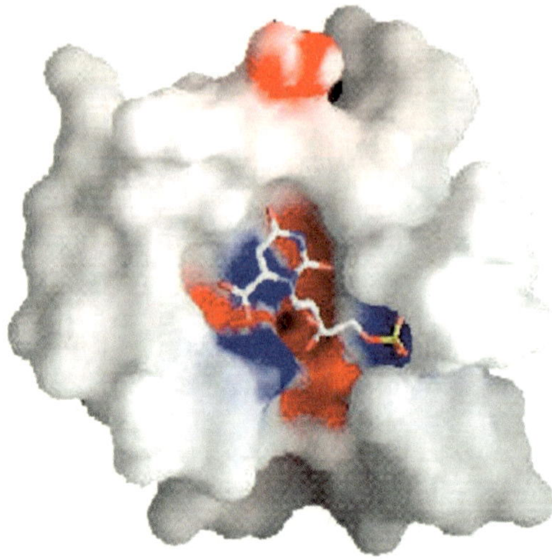

Fig. 8 Electrostatic potentials over the van der Waals surface of the active site environment for the reactant state. Electrostatic potentials are represented in *red* for negative potentials, in *blue* for positive potentials , and in *white* for neutral potentials. We thank Ning Wu for making this figure using GRASP

sults in the formation of a third hydrogen bond to the 2′-hydoxyl group from Lys72. This is absent in the X-ray structure.

The interactions of the ribosyl 2′-hydoxyl on catalysis has been implicated in experiments with a deoxy-OMP substrate (see below) [60]. The orotate carboxylate group does not form an ion pair either with Lys72 or Lys42 because the former would have brought three carboxylate groups to close proximity causing repulsions to dominate. Lys42 is hydrogen bonded to the ribosyl 3′-hydroxyl group in the opposite site (Fig. 5). It is this delicate balance of electrostatic and hydrophobic environment in the orotate binding site that creates a highly reactive conformational substate of the enzyme that facilitates the decarboxylation reaction. Clearly, any mutation of these charged residues will disrupt this critical electrostatic balance in the active site.

This mechanism is consistent with the computational results listed in Fig. 3 for the Asp70Ala mutant. The lowering of 7.5 kcal/mol in the free energy barrier in the Asp70Ala mutant is probably due to electrostatic stabilization in the enzyme active site. The inclusion of a chloride ion in the location normally occupied by the Asp70 residue partially restores the electrostatic features of the wild-type enzyme and introduces electrostatic stress due to interactions between this ion and the substrate carboxylate ion. This electrostatic stress induces conformational changes with increased protein distortion energy in the Michaelis complex, contributing a reduction of the activation barrier by about 10 kcal/mol [21]. Interestingly, the Asp70Ala mutant does not show a significant increase in substrate binding affinity because the active site is rescued by a small ion to maintain the electrostatic and hydrogen bonding network. The wild-type enzyme introduces additional cooperative interactions by covalently connecting the anionic charge to the Asp residue, further lowering the free energy of activation by 6.2 kcal/mol.

It is important to distinguish clearly between the notions of *observed* and *intrinsic* free energies of binding. For similar values of observed free energy of binding by wild-type and mutant enzymes, their origins in protein distortion and intrinsic binding can be very different. The idea that a reactant state destabilization mechanism requires very strong *intrinsic* binding is nonetheless consistent with a much smaller *observed* binding free energy. This is because the observed binding energy is a sum of the intrinsic binding energy and the protein distortion energy. Thus, one should not expect that a reactant state destabilization mechanism necessarily requires a very large observed binding affinity for the substrate [28].

The involvement of protein conformational energy in ODCase catalysis is reflected by experimental studies cited by Miller and Wolfenden [56]. They found that ~15 kcal/mol can result from cooperative interactions between the phosphoribosyl group and the orotate group simply by disconnecting these two groups [56]. If the enzyme environment were rigid or if interactions within the protein did not contribute to the observed binding energy and catalytic reaction barrier, there would have not been such large cooperative effects.

One test of a proposed catalytic mechanism is its ability to be applied to the results of site-directed mutagenesis. In the present case, the effect of altering any one of the four charged residues (Asp70, Asp75b, Lys42, and Lys72) can be understood because these changes affect the stability and interactions in the active site so dramatically that the enzyme's activity is lost [20]. Furthermore, the catalytic activity could be partially restored by appropriate small ions. What is intriguing is that mutations in the phosphate binding pocket have large effects on k_{cat} but produce only small changes in K_M. For example, the mutation of Tyr217Ala (yeast ODCase numbering in the mutation study) increases the barrier height by 4.7 kcal/mol, whereas the change in binding affinity is about 1 kcal/mol [14]. In fact, Tyr217 is at least 8 Å from the reactive orotate group, making it difficult to argue for a transition state stabilization mechanism. On the basis of enzyme reactant state destabilization, the Tyr217Ala mutation reduces hydrogen bonding interactions with the phosphate group. At the same time, it also induces smaller protein distortion than in the wild-type enzyme. Consequently, these two factors cancel, giving rise to a small change in K_M. However, since the enzyme is less distorted in the reactant state, there is less energy released from protein relaxation during the chemical transformation, giving rise to greater reduction in the observed k_{cat} value.

It was predicted that the rate would be decreased if the OMP substrate is replaced by 2′-deoxyOMP substrate because the hydroxyl group plays an important role in interacting with Lys72 to stabilize the distorted reactant state [16]. This was confirmed by the work of Wolfenden and coworkers who showed that decarboxylation barrier of 2′-deoxyOMP is 4.6 kcal/mol higher than that of OMP with small binding contributions [60]. In contrast, in a TS stabilization mechanism this alteration would be expected to increase K_M significantly with little effect on k_{cat} as the loss of hydrogen bonding interactions remains the same both in the reactant and transition state.

6 Conclusions

In this article, we present a reactant state destabilization mechanism that arises from enzyme conformational distortion induced upon binding of the substrate. This can account for the remarkable rate increase in the decarboxylaiton reaction catalyzed by ODCase. Large protein distortions can result from unfavorable electrostatic interactions between the enzyme and a small part of the substrate, although the overall binding interaction is still favorable. As the decarboxylation of OMP proceeds in ODCase, the protein conformation becomes less distorted at the transition state in response to the charge reorganization of the substrate, releasing the reactant state electrostatic stress. Although the reactant is not particularly more stabilized at the transition state by the enzyme than is the reactant state, the change of protein conformational energy helps to reduce the overall free energy of activation. This mechanism is consistent with observations that mutations away from the active site can have large effects on k_{cat} but not on K_M because of the compensating factors of protein distortion and intrinsic substrate binding interactions.

Acknowledgements We thank the NIH for support for the research at the University of Minnesota. RK thanks NSERC Canada for support through a Discovery Grant.

References

1. Radzicka A, Wolfenden R (1995) Science (Washington DC) 267:90–93
2. Lad C, Williams NH, Wolfenden R (2003) Proc Natl Acad Sci USA 100:5607–5610
3. Jencks WP (1975) Adv Enzymol Relat Areas Mol Biol 43:219–410
4. Pauling L (1946) Chem Eng News 24:1375
5. Schowen RL (1978) In: Gandour RD, Schowen RL (eds) Transition states of biochemical processes. Plenum, New York, pp 77–114

6. Karplus M, McCammon JA (2002) Nature Struc Biol 9:646–652
7. Colonna-Cesari F, Perahia D, Karplus M, Eklund H, Branden CI, Tapia O (1986) J Biol Chem 261:15273–15278
8. Rozovsky S, Jogl G, Tong L, McDermott AE (2001) J Mol Biol 310:271–280
9. Sawaya MR, Kraut J (1997) Biochemistry Field 36:586–603
10. McMillan FM, Cahoon M, White A, Hedstrom L, Petsko GA, Ringe D (2000) Biochemistry 39:4533–4542
11. Stroud RM, Finer-Moore JS (2003) Biochemistry 42:239–247
12. Finer-Moore JS, Santi DV, Stroud RM (2003) Biochemistry 42:248–256
13. Wagner CR, Huang Z, Singleton SF, Benkovic SJ (1995) Biochemistry 34:15671–15680
14. Miller BG, Snider MJ, Short SA, Wolfenden R (2000) Biochemistry 39:8113–8118
15. Appleby TC, Kinsland C, Begley TP, Ealick SE (2000) Proc Natl Acad Sci USA 97:2005–2010
16. Wu N, Mo Y, Gao J, Pai EF (2000) Proc Natl Acad Sci USA 97:2017–2022
17. Miller BG, Hassell AM, Wolfenden R, Milburn MV, Short SA (2000) Proc Natl Acad Sci USA 97:2011–2016
18. Harris P, Poulsen J-CN, Jensen KF, Larsen S (2000) Biochemistry 39:4217–4224
19. Traut TW, Temple BRS (2000) J Biol Chem 275:28675–28681
20. Miller BG, Snider MJ, Wolfenden R, Short SA (2001) J Biol Chem 276:15174–15176
21. Wu N, Gillon W, Pai EF (2002) Biochemistry 41:4002–4011
22. Wu N, Pai EF (2002) J Biol Chem 277:28080–28087
23. Feng WY, Austin TJ, Chew F, Gronert S, Wu W (2000) Biochemistry 39:1778–1783
24. Lee JK, Houk KN (1997) Science (Washington DC) 276:942–945
25. Silverman RB, Groziak MP (1982) J Am Chem Soc 104:6434–6439
26. Lee T-S, Chong LT, Chodera JD, Kollman PA (2001) J Am Chem Soc 123:12837–12848
27. Hur S, Bruice TC (2002) Proc Natl Acad Sci USA 99:9668–9673
28. Warshel A, Strajbl M, Villa J, Florian J (2002) Biochemistry 39:14728–14738
29. Allen KN, Lavie A, Petsko GA, Ringe D (1995) Biochemistry 34:3742–3749
30. Lavie A, Allen KN, Petsko GA, Ringe D (1994) Biochemistry 33:5469–5480
31. Aaqvist J, Warshel A (1993) Chem Rev (Washington DC) 93:2523–2544
32. Warshel A, Florian J (1998) Proc Natl Acad Sci USA 95:5950–5955
33. Garcia-Viloca M, Gao J, Karplus M, Truhlar DG (2004) Science (in press)
34. Warshel A, Florian J, Strajbl M, Villa J (2001) Chem Bio Chem 2:109–111
35. Yu EW, Koshland DE Jr (2001) Proc Natl Acad Sci USA 98:9517–9520
36. Guo H, Cui Q, Lipscomb WN, Karplus M (2003) Angew Chem, Int Edn 42:1508–1511
37. Hur S, Bruice TC (2003) J Am Chem Soc 125:5964–5972
38. Guimaraes CRW, Repasky MP, Chandrasekhar J, Tirado-Rives J, Jorgensen WL (2003) J Am Chem Soc 125:6892–6899
39. Strajbl M, Shurki A, Kato M, Warshel A (2003) J Am Chem Soc 125:10228–10237
40. Koshland DE Jr (1960) Adv Enzymol 22:45–97
41. Warshel A, Levitt M (1976) J Mol Biol 103:227–249
42. Gao J, Xia X (1992) Science 258:631–635
43. Bash PA, Field MJ, Karplus M (1987) J Am Chem Soc 109:8092–8094
44. Gao J (1992) J Phys Chem 96:537–540
45. Field MJ, Bash PA, Karplus M (1990) J Comput Chem 11:700–733
46. Gao J (1994) ACS Symp Ser 569:8–21
47. Gao J, Truhlar DG (2002) Ann Rev Phys Chem 53:467–505
48. Gao J, Pavelites JJ (1992) J Am Chem Soc 114:1912–1914
49. Gao J (1995) J Am Chem Soc 117:8600–8607
50. Gao D, Pan Y-K (1999) J Org Chem 64:4492–4501

51. Gao J (1994) Proc Ind Acad Sci 106:507–519
52. Orozco M, Luque FJ, Habibollahzadeh D, Gao J (1995) J Chem Phys 102:6145–6152
53. Gao J (1997) J Comput Chem 18:1062–1971
54. Kollman P (1993) Chem Rev (Washington DC) 93:2395–2417
55. Rajamani R, Naidoo K, Gao J (2003) J Computational Chem (in press)
56. Miller BG, Wolfenden R (2002) Ann Rev Biochem 71:847–885
57. Dewar MJS, Zoebisch EG, Healy EF, Stewart JJP (1985) J Am Chem Soc 107:3902–3909
58. Jorgensen WL (1989) Acc Chem Res 22:184–189
59. Alhambra C, Wu L, Zhang Z-Y, Gao J (1998) J Am Chem Soc 120:3858–3866
60. Miller BG, Butterfoss GL, Short SA, Wolfenden R (2001) Biochemistry 40:6227–6232

Author Index Volumes 201–238

Author Index Vols. 26–50 see Vol. 50
Author Index Vols. 51–100 see Vol. 100
Author Index Vols. 101–150 see Vol. 150
Author Index Vols. 151–200 see Vol. 200

The volume numbers are printed in italics

Subject Index

Printing: Saladruck, Berlin
Binding: Stein+Lehmann, Berlin